Induced Seismic Events

Edited by
Peter Knoll
Georg Kowalle

1996

Birkhäuser Verlag
Basel · Boston · Berlin

Reprint from Pure and Applied Geophysics
(PAGEOPH), Volume 147 (1996), No. 2

The Editors:

Professor Dr. Peter Knoll
Dr. Georg Kowalle
GTU Ingenieurbüro Knoll
Potsdamer Str. 18 A
D-14513 Teltow
Germany

A CIP catalogue record for this book is available from the Library of Congress, Washington D.C., USA

Deutsche Bibliothek Cataloging-in-Publication Data

Induced seismic events / ed. by Peter Knoll; Georg Kowalle. –
Basel ; Boston ; Berlin : Birkhäuser, 1996
  Aus: Pure and applied geophysics ; Vol. 147. 1996
ISBN 978-3-7643-5454-2   ISBN 978-3-0348-9204-9 (eBook)
DOI 10.1007/978-3-0348-9204-9
NE: Knoll, Peter [Hrsg.]

© 1996 Birkhäuser Verlag, P.O. Box 133, CH-4010 Basel, Switzerland
Printed on acid-free paper produced from chlorine-free pulp TCF ∞
ISBN 978-3-7643-5454-1

987654321

Contents

PAGEOPH, Vol. 147, No. 2 (1996)

0033–4553/96/020205–02$1.50 + 0.20/0
© 1996 Birkhäuser Verlag, Basel

Preface

In Europe studies of induced seismic events have a long tradition. They are of high scientific and economic importance due to the extensive mining activities and the high density of population within this region.

Over recent decades the European Seismological Commission (ESC) initiated and stimulated corresponding research. During the XXIV General Assembly (September 19–24, 1994) in Athens a special symposium was organized by the Working Group "Induced Seismic Events" of the Sub-Commission E "Earthquake Prediction Research". The meeting was chaired by P. Knoll and S. Gibowicz. Eighteen papers were presented orally and seven posters were included in the scope of the symposium. The authors were invited to present contributions to this special issue of Pure and Applied Geophysics. Additionally two Chinese papers could be included in this volume.

One focus of contributions concentrates on induced seismicity in the mining districts of Europe, especially in Middle and Eastern Europe. Papers were presented with description of induced seismicity in Austria, the Upper Silesia mining district, Bulgaria and at Kola Peninsula. Techniques of analysing mining-induced events were presented, indicating the strong efforts in studying the technological prerequisites and tectonic sources of those events. These papers must be seen in direct connection with papers presenting observations and results from South Africa and Canada. Especially interesting new techniques for analyzing induced seismic events are included.

The second main topic of investigations concentrates on reservoir-induced seismicity at the Toktogul reservoir in Kirghizia and selected reservoirs in China and extraction-induced seismic events in the Gazli-region in Uzbekistan. These papers illustrated the strong interaction of the tectonically determined stress field with local stress changes due to changes of crustal loading.

Theoretical and laboratory studies are presented dealing with the description of the rupture process of induced events. These techniques have been applied to explain the focal processes of induced seismic events.

The editors wish to express thanks to all of the contributors for their skilled work and for their presentation of the papers to facilitate publication within the frame of this volume.

Special thanks are directed to the reviewers Michael Baumbach, Josef Dubiński, Bruno Feignier, Siegfried Franck, Don J. Gendzwill, Slawomir Gibowicz,

Helmut Grosser, Harsh K. Gupta, Eckart Hurtig, Andzej Kijko, Petr Konecny, B. K. Rastogi, V. Rudajev, Fritz Rummel, Ulfert Seipold, Genady Sobolev, C. Srinivasan, Cezar Trifu and Manfred Will for their constructive criticism, which produced an important impact towards improving the quality of the submitted articles.

P. Knoll   G. Kowalle
GTU Ingenieurbüro Knoll & GeoDyn GmbH
Potsdamer Str. 18A, D-14513 Teltow, Germany

PAGEOPH, Vol. 147, No. 2 (1996)

0033–4553/96/020207–10$1.50 + 0.20/0

The Mechanism of Mine-collapse Deduced from Seismic Observations

WOLFGANG A. LENHARDT[1] and CHRISTIANE PASCHER[1]

Abstract—On May 2, 1993 more than 200 seismic events from an underground mine in Tyrol/Austria were recorded with short-period seismometers of a local seismic network which was introduced in the late 1980s to monitor the tectonic seismicity in Tyrol in greater detail. The cause of this series of mining-associated events has become the subject of intensive investigations—as it was associated with a subsidence affecting an area of 10.000 m². Underground observations revealed a number of discontinuities along which the rock mass was able to move. Seismic recordings of the close-by seismic stations revealed two types of mechanisms: One mechanism seems to be associated with pure block-sliding along several discontinuities, while other signals indicate additional collapse. The consideration and combination of several seismological principles made possible the construction of a model of the mine collapse.

Key words: Mine collapse, seismic signals, long-period wavelet, first motions, collapse model, possible cause.

1. Introduction

On Sunday, May 2, 1993 a sequence of seismic events was recorded with short-period seismometers from the area of Schwaz, a town in Tyrol/Austria.

The Department of Geophysics of the Central Institute for Meteorology and Geodynamics in Vienna (Austria) maintains four seismic stations in Tyrol, equipped with Geotech S13 short-period seismometers which can be directly accessed from Vienna. Two of them (station Wattenberg 'WTTA' and Walderalm 'WATA') are situated 12 km from Schwaz (Fig. 1). It turned out that the observed seismic signals were related to a partial collapse of dolomite-mine workings near this town.

Mining has a long tradition in Schwaz, reaching its peak in production during the 16th century when silver was recovered from numerous galleries in the mining district of 'Falkenstein' near Schwaz (EGG *et al.*, 1986). Today's production concentrates on underground dolomite mining. Mine workings extend over 200 m in a vertical direction, and finally reach a diameter of approximately 80 m (see also Fig. 7).

[1] Central Institute for Meteorology and Geodynamics, Department of Geophysics, Hohe Warfe 38, A-1190, Vienna, Austria.

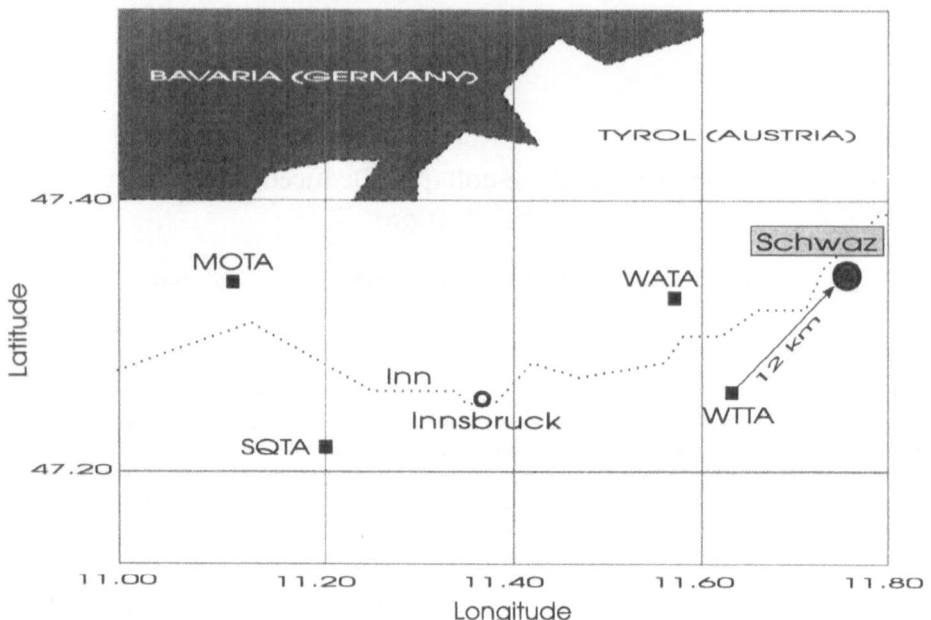

Figure 1
Seismic stations in Tyrol/Austria (solid squares) and the mining district of Schwaz.

The main shock registered a magnitude of M 1.9, which was even observed in Vienna as well as by foreign seismological stations. It constituted the last event of a sequence of events which was associated with a massive subsidence, reaching 8.5 m in places (CZUBIK, 1993), 340 m above the mine workings. The entire process took two hours and affected a surface area of 80 m × 150 m, thus roughly delineated the plan view of one of the involved underground caverns. Underground observations revealed a number of discontinuities along which the rock mass was able to move. The total rock mass, which was involved in the process, amounted to approximately half a million cubic meters.

2. Seismic Observations

2.1. Time Sequence of Collapse

The seismic activity began on May 2, 1993 at 1 h 15 UTC (=local time − 2 hours) with minor events. At 2 h 15 the first larger event was recorded, registering a magnitude of M 1.4. The next stronger shock occurred at 2 h 57 (M 1.7), and shortly thereafter the main event (M 1.9) at 2 h 59.

The cumulative frequency-magnitude relationship from this 2-hour long sequence can be described by

$$\log N = 1.93 - 0.95\, M, \quad \text{valid for } -0.7 \leq M \leq 2$$

which indicates that more than 390 events ('N') of $M \geq -0.7$ were observed during this period of time. At least half of them could be further evaluated in terms of first motions and wavelets. The remaining signals could not be evaluated, mainly due to insufficient separation in time.

Ths slope (0.95) of this distribution, which is commonly referred to as the 'b-value', obviously does not differ from aftershock sequences of tectonic earthquakes ($b = 0.5-1.0$), thus indicating that the size distribution of the involved slip-planes seems to follow fractal laws, similar to those of earthquake processes (AKI, 1981).

However, two types of seismic signals could be distinguished, which are uncommon in earthquakes of tectonic origin. These signals were also observed in subsequent seismic activities in November/December 1993 and April 1994, which indicated that the process of collapse has not been fully completed.

2.2. Two Kinds of Seismic Signals

Based on the shape of the signals, it was possible to distinguish 'collapse' events from tectonic earthquakes. The first kind of signals (1 in Fig. 2) exhibits a long-period wavelet, starting approximately 3 seconds after the arrival of the

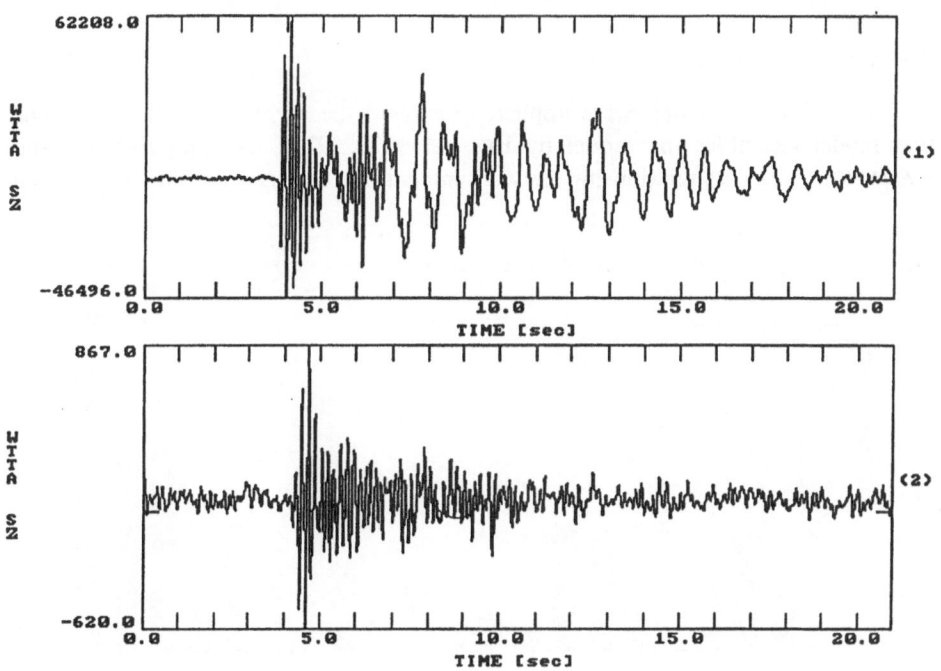

Figure 2
Seismic signals with long (1) and short-period (2) wavelet.

compressional wave. This long-period wavelet is missing in the second type (2 in Fig. 2) of seismic signals. Sometimes both types of signals showed a 'compressional' (movement away from the source), and sometimes a 'dilatational' (movement towards the source) first arrival at the very same station. As it can be reasonably assumed that both signals originated in the mine and were associated with the collapse, we are left with the question, why could two obviously different signals be observed?

A combination of a slip and collapse mechanism seems to explain both types of signals. Type 1 would involve the slip of a rock mass along a plane of weakness, including a free fall of the rock mass, which terminates with an impact pulse, thus creating the long-periodic wavelet. Type 2 events lack these onsets, hence they seem to be caused by pure slip movements.

The process obviously started with events of the 1-type (Fig. 3), indicating the collapse of the roof. At a later stage, events of both types can be observed. At the end of the seismic sequence mainly type 1-events were observed again.

However, the erratic presence of compressional and dilatational onsets indicates the possible involvement of more than one slip-plane, with an opposing sense of movement. Only from the point of mining geometry does this situation actually apply to the real situation. The cavern under consideration is in fact surrounded by more than two discontinuities (see also Fig. 7).

2.3. Seismological Models

Three different models were applied to explain the involved mechanisms. The first model resembles the model by BRUNE (1970, 1971), including HANKS and KANAMORI's (1979) relationship between magnitude and seismic moment, which

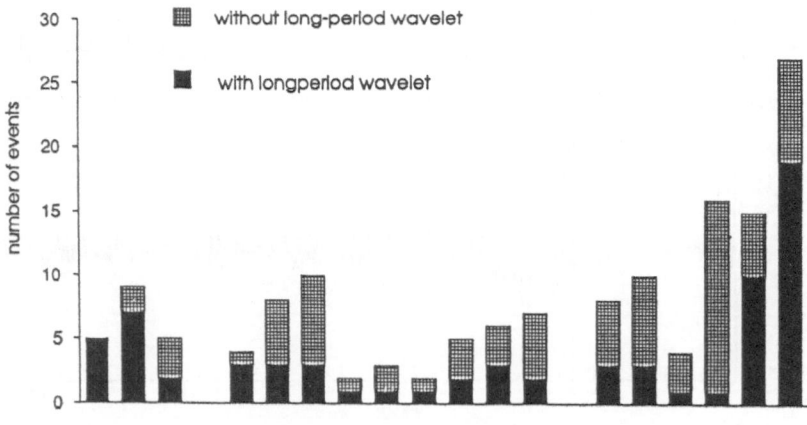

Figure 3
Frequency of type 1 (solid fill) and type 2 wavelets (hatched fill) during the main collapse.

implies a stress drop $\Delta\sigma$ assumption of 0.01% of the shear modulus, when used in connection with GUTENBERG and RICHTER's (1956) relationship between seismic energy and magnitude. The Brune model allows for estimation of the size of the involved rupture area:

$$A = M_0/(d * G)$$

with

A —rupture area (m^2)
M_0—seismic moment (Nm)
d—displacement (m)
G—shear modulus (33 GPa).

Hence, a magnitude $M2$-event is related to a source radius of 35 m, if the displacement is kept at 0.01 m. Larger displacements automatically result in smaller source-radii; hence the size of contact-planes between dislodged blocks and the surrounding rock mass decreases, which ultimately leads to smaller involved volumes of the rock masses.

The second model allows for calculation of the involved volume of rock V in m^3, which moved purely gravitational. The volume can be estimated by relating the potential energy E_{pot} and the released seismic energy E_s in Joule (GUTENBERG and RICHTER, 1956) by

$$E_s = 10^{(1.5 * M + 4.8)}$$

$$E_{pot} = V * \rho * g * h$$

$$E_s = E_{pot} * \eta$$

hence

$$V = E_s/(\rho * g * h * \eta)$$

with

ρ—density (2850 kg/m^3)
g—acceleration of gravity (9.81 m/s^2)
h—height of free fall (m)
η—seismic efficiency (E_s/E_{pot}).

Given a height of 25 m and a seismic efficiency η of 0.001 (thus only 0.1% of the potential energy is actually converted into seismic energy), the main event with a magnitude of M 1.9 relates to a volume of 63.000 m^3, which is insufficient when compared with the existing cavern of 500.000 m^3.

The third model regards slip along planes and collapse. Each block is treated as if it would have been sliding along its cylindrical surface for 0.025% of its height. The rupture area A is determined by the radius of the cylinder and its height

$$A = 2 * \pi * r * H$$

with

r—radius of the cylinder (m)
H—height of the cylinder (m).

After sliding, the block is subjected to a free fall. A small volume of rock, with a radius of 4 m and a height of 2 m, would cause an event of $M0$ coupled with a dropping height of 25 m and a seismic efficiency of 1%. A larger block, with a radius of 45 m and a height of 20 m, would cause an event of $M2$. The involved volume totals 127.000 m³ in the latter case, which is still lower than the available volume of the cavern. It should be noted, however, that a lower seismic efficiency automatically leads to larger volumes in this approach.

Although many assumptions had to be made (consistent slip and stress drop, seismic efficiency of slip event equals the impact event), some conclusions can be drawn from the above estimates:

1) sliding in conjunction with collapse as well as pure sliding dominate the model—and not pure collapse.
2) seismic efficiencies seem to range between 0.1% and 1%.

It should be noted, that the results from the three simple models presumedly serve only as rough estimates, since they necessarily include numerous assumptions.

3. Interpretation

Already the time of collapse, early on a Sunday morning indicates that the collapse was not directly triggered by mining activities, such as production blasts, as this is generally the case (see also LENHARDT, 1992). Furthermore, seismicity in November/December 1993 and in April 1994 did not correlate with blasting time either. Figure 4 shows the diurnal distribution of these seismic events, in which the sequence of May 2, 1993 has been excluded for clarity.

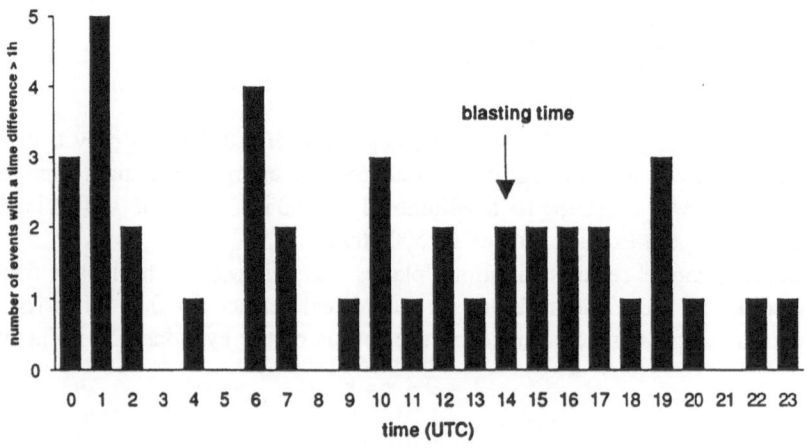

Figure 4
Diurnal distribution of seismic activity.

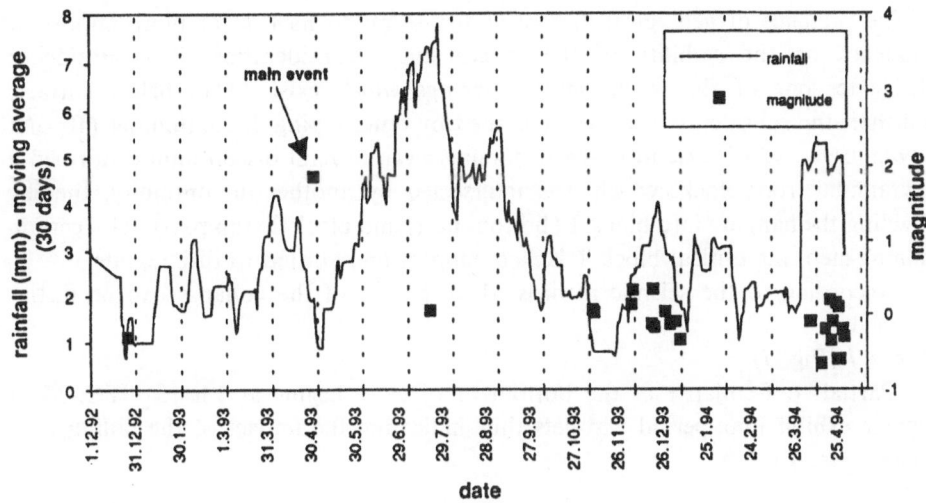

Figure 5
Rain- and snowfall (30 days-moving average).

However, meteorological influences cannot be completely ruled out. As can be seen in Figures 5 and 6, the seismicity in November/December 1993 and April 1994 correlated with peaks of snow- or rainfall. This fact does not apply to the time of the main collapse, which occurred during a period of thaw, after a two-month increase of temperature.

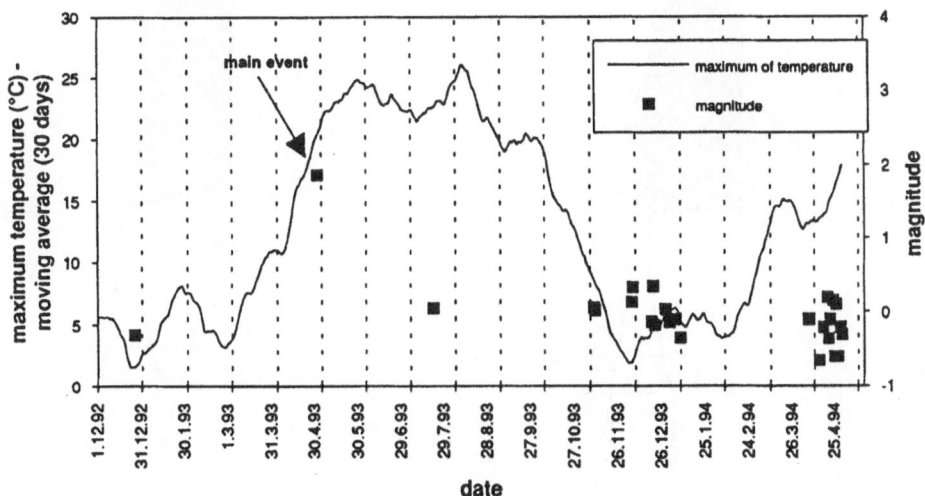

Figure 6
Temperature (30 days-moving average).

The presence of geological discontinuities seems to have a far more important influence on the stability of the cavern under consideration. The subsidence delineates one of the recent mine workings which exist 300 m below surface. Further, the subsidence was clearly limited by outcropping discontinuities (BAUER et al., 1993), which extend down to mining level. Several discontinuities formed a 'pyramidal' roof-block, which was intersected by another discontinuity, thereby dividing the hanging into block 1 (bottom part) and block 2 (top part). At a certain mining-step, the critical block 1 lacked support and started to disintegrate.

According to the seismic records, three phases of the collapse can be distinguished:

Phase 1 (Fig. 7)

Partial roof-collapse at the bottom of block 1 begins at 1 h 15 UTC. Most signals exhibit long-period wavelets thus indicating the impact of the falling rock masses (see Fig. 3).

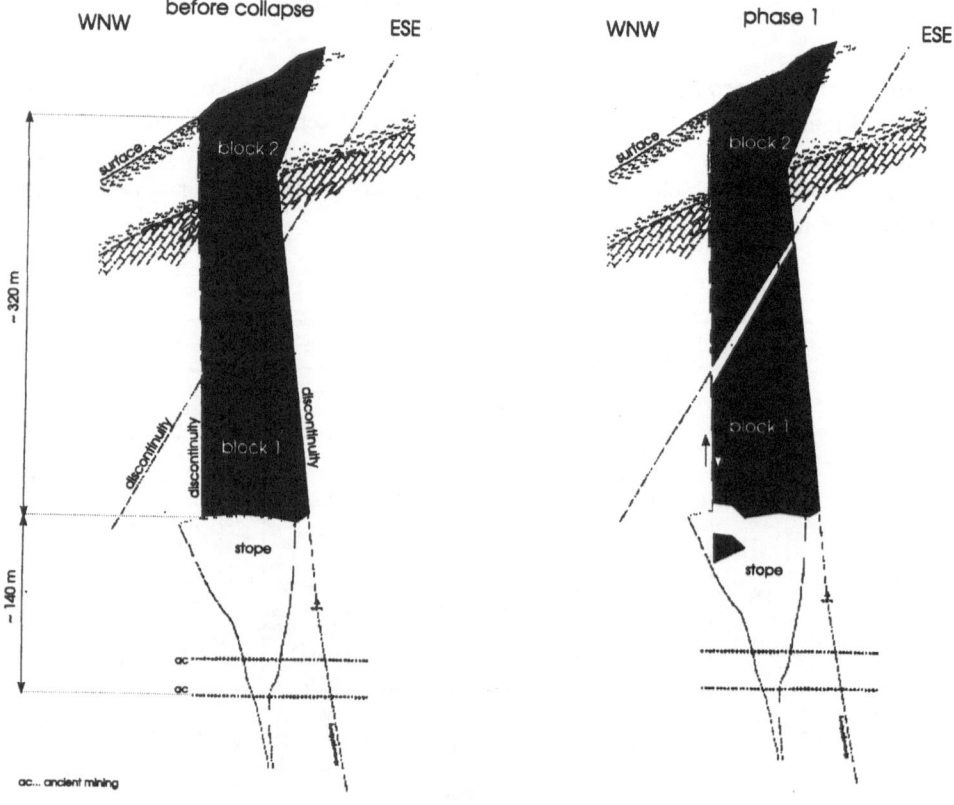

Figure 7
Initial state and proposed mechanism during phase 1.

Phase 2 (Fig. 8)

Due to changed stress conditions, block 1 mainly moves along the existing—almost vertically orientated—discontinuities. This process seems to have started at 1 h 45. Erratic compressional and dilatational onsets at one and the same seismic station indicate the involvement of more than one slip-plane. Finally, block 1 disintegrates. Block 2 starts to move.

Phase 3 (Fig. 8)

Since block 1 has disintegrated, block 2 is able to move due to lack of vertical support. Finally, at 2 h 59, the main part of block 2 seems to have moved at once, thus emitting most of the seismic energy (with long-period wavelets) and causing the subsidence on surface.

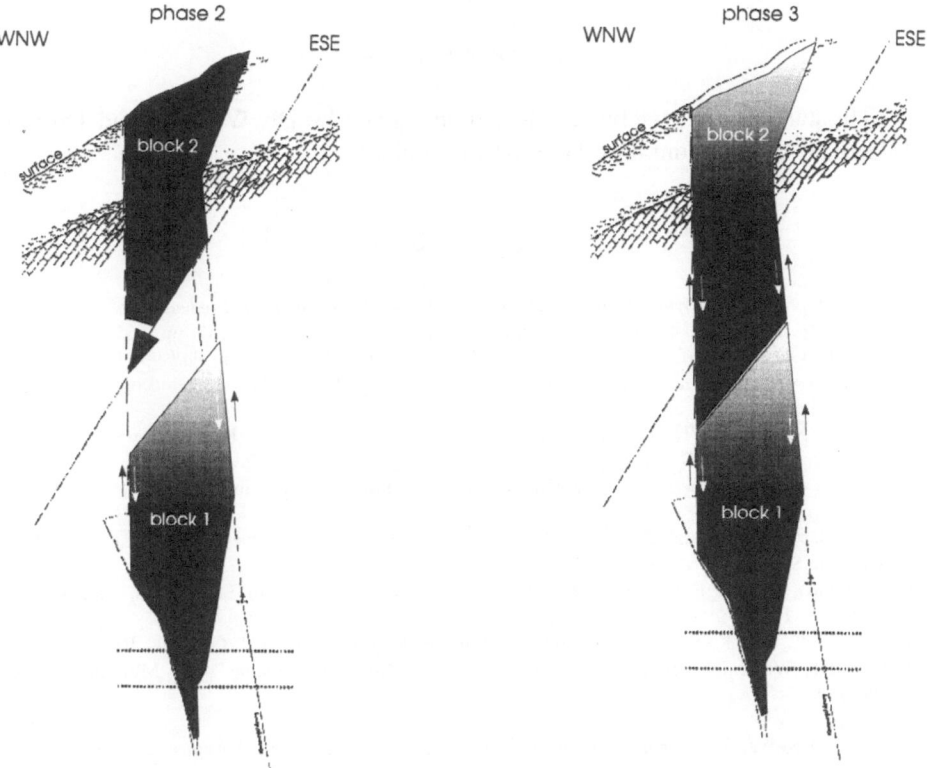

Figure 8
Proposed mechanisms during phase 2 and phase 3.

4. Summary

Seismic recordings of nearby seismic stations revealed two types of mechanisms during the collapse: One mechanism seems to be associated with pure block-sliding along several discontinuties, while other signals indicate collapse events.

Three phases of the collapse can be recognized from seismic observations. The first phase seems to be a result of slip movement along one of the adjacent discontinuities and collapse of the roof. During the second phase the lower part of the roof block collapses, fills the mine workings, and the higher part of the block starts to move. The third phase constitutes the final part of the process during which the subsidence on surface took place.

Not only the atypical history of seismic events—numerous foreshocks, culminating in the main shock at the end of the collapse—attracts attention, but also the fact that minor seismic activity continues, thus indicating remaining cavities in the mine workings under consideration.

Acknowledgement

The authors are grateful for the support given by Dr. Wöbking and Dipl.Ing. Ludescher of the Montanwerke Brixlegg Limited.

References

AKI, K. *A probabilistic synthesis of precursory phenomena.* In *Earthquake Prediction, an International Review.* M. Ewing Ser. 4 (Simpson, D., and Richards, P., eds.) (AGU, 1981) pp. 566–574.

BAUER, C. G., OHLBOTH, S., WENGER, H., and RIESER, B. (1993), *Bergbau Falkenstein der Montanwerke Brixlegg GmbH*, Geologic-tectonic Mapping and Documentation, Minccon Geo GmbH, unpublished.

BRUNE, J. N. (1970/1971), *Tectonic Stress and Spectra of Seismic Shear Waves from Earthquakes*, J. Geophys. Res. *75*, 4997–5009, Correction in J. Geophys. Res. *76*, 5002.

CZUBIK, E. (1993), *Vermessung des Bergbaus Schwaz.* Requested Investigation by Montanwerke Brixlegg Ltd., unpublished.

EGG, E., GSTREIN, P., and STERNAD, H. (1986), *Stadtbuch Schwaz. Natur-Bergbau-Geschichte.* Stadtgemeinde Schwaz, Tyrol, Austria.

GUTENBERG, B., and RICHTER, C. F. (1956), *Magnitude and Energy of Earthqiakes*, Ann. Geophis. (Rome) *9*, 1–115.

HANKS, T. C., and KANAMORI, H. (1979), *A Moment Magnitude Scale*, J. Geophys. Res. 2348–2350.

LENHARDT, W. A. (1992), *Seismicity Associated with Deep-level Mining*, Acta Montana *A2* (88), 179–192.

(Received December 22, 1994, revised August 23, 1995, accepted August 24, 1995)

PAGEOPH, Vol. 147, No. 2 (1996)

0033–4553/96/020217–22$1.50 + 0.20/0

Tectonic Analysis of Mine Tremor Mechanisms from the Upper Silesian Coal Basin

Grzegorz Sagan[1], Lesław Teper[1] and Waclaw M. Zuberek[1]

Abstract—Fault network of the Upper Silesian Coal Basin (USCB) is built of sets of strike-slip, oblique-slip and dip-slip faults. It is a typical product of force couple which acts evenly with the parallel of latitude, causing horizontal and anti-clockwise movement of rock-mass. Earlier research of focal mechanisms of mine tremors, using a standard fault plane solution, has shown that some events are related to tectonic directions in main structural units of the USCB. An attempt was undertaken to analyze the records of mine tremors from the period 1992–1994 in the selected coal fields. The digital records of about 200 mine tremors with energy larger than 1×10^4 J ($M_L > 1.23$) were analyzed with SMT software for seismic moment tensor inversion. The decomposition of seismic moment tensor of mine tremors was segmented into isotropic (I) part, compensated linear vector dipole (CLVD) part and double-couple (DC) part. The DC part is prevalent (up to 70%) in the majority of quakes from the central region of the USCB. A group of mine tremors with large I element (up to 50%) can also be observed. The spatial orientation of the fault and auxiliary planes were obtained from the computations for the seismic moment DC part. Study of the DC part of the seismic moment tensor made it possible for us to separate the group of events which might be acknowledged to have their origin in unstable energy release on surfaces of faults forming a regional structural pattern. The possible influence of the Cainozoic tectonic history of the USCB on the recent shape of stress field is discussed.

Key words: Mining induced seismicity, seismic moment tensor, focal mechanism, Upper Silesia, fault tectonics.

Introduction

In recent times calculation of the seismic moment tensor has replaced the classic approach to the solution of the mine tremors mechanism. In comparison with the classic fault plane solution, it gives us the opportunity to consider components other than the shear one. The seismic moment tensor is not a common method in Polish coal mines, due mainly to the poor quality of seismic recordings. Although the method is known to the mining seismicity community (e.g., Sileny, 1989; McGarr, 1992a,b; Feignier and Young, 1992), only a few papers describe such studies in Poland, e.g., in Polish Copper Mines (Wiejacz, 1991; Gibowicz, 1992, 1993) and the Upper Silesian Coal Basin-USCB (Wiejacz, 1994).

The algorithm used in this research is the amplitude inversion in time domain, computed by the latest version of the SMT program elaborated by Wiejacz (1994). The theoretical background of the algorithm used was also presented in Wiejacz's Ph.D. thesis (1991) and in the paper by Gibowicz (1993) or Gibowicz and Kijko

---

[1] Silesian University, Faculty of Earth Sciences, ul. Bedzinska 60, 41–200 Sosnowiec, Poland.

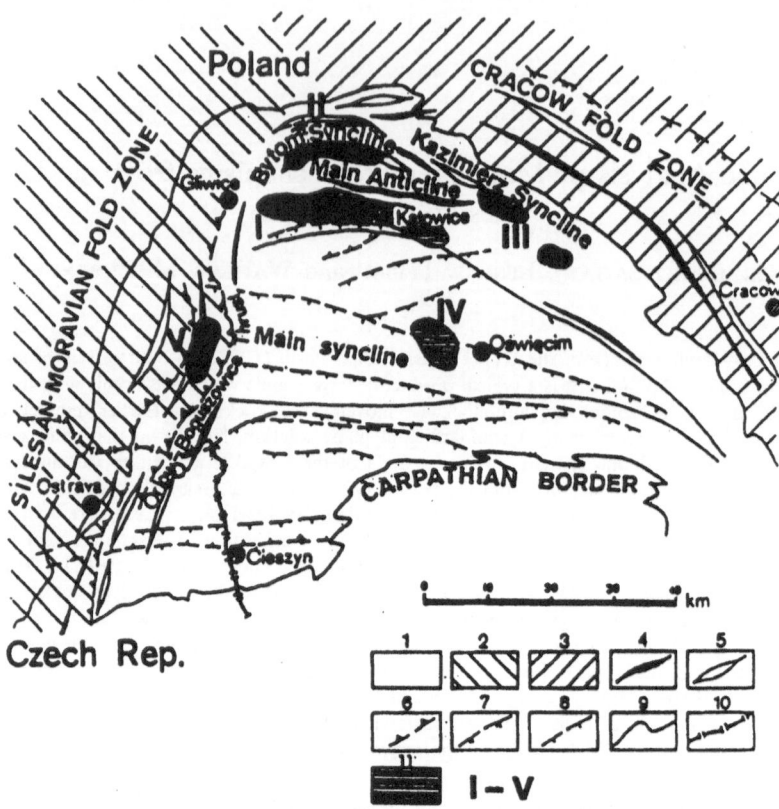

Figure 1

The sketch of geological structure of the Upper Silesian Coal Basin with the Wujek coal field localizations. 1. Zone of block tectonics; 2. zone of fold tectonics; 3. zone of fold-block tectonics; 4. anticlines; 5. synclines; 6. thrusts; 7. main faults; 8. main faults of Alpine age or reactivated ones; 9. border of the USCB; 10. state border; 11. seismic areas; Wujek mine area (cf., Fig. 2) is shaded.

(1993), which included the part pertaining to numerical tests of the calculation program. The above-mentioned latest version of software was used in this investigation of mining tremors in the Wujek and Ziemowit mines (WIEJACZ, 1994). The moment-tensor inversion was performed in time domain using the first motion amplitudes and signs of P and sometimes SV waves. The relatively simple isotropic geological model was assumed, and the source location, waves velocity, rock-mass density, and the frequency characteristics of the mine underground network were used as input data. The determination of the moment tensor is performed in four steps. In the first one the displacements are transformed into the radial and tangential components. In the second step the deconvolution of the seismic signal due to hardware characteristics is performed, as well as the removal of high frequency noise and the correction of amplitudes due to the epicentral distance of

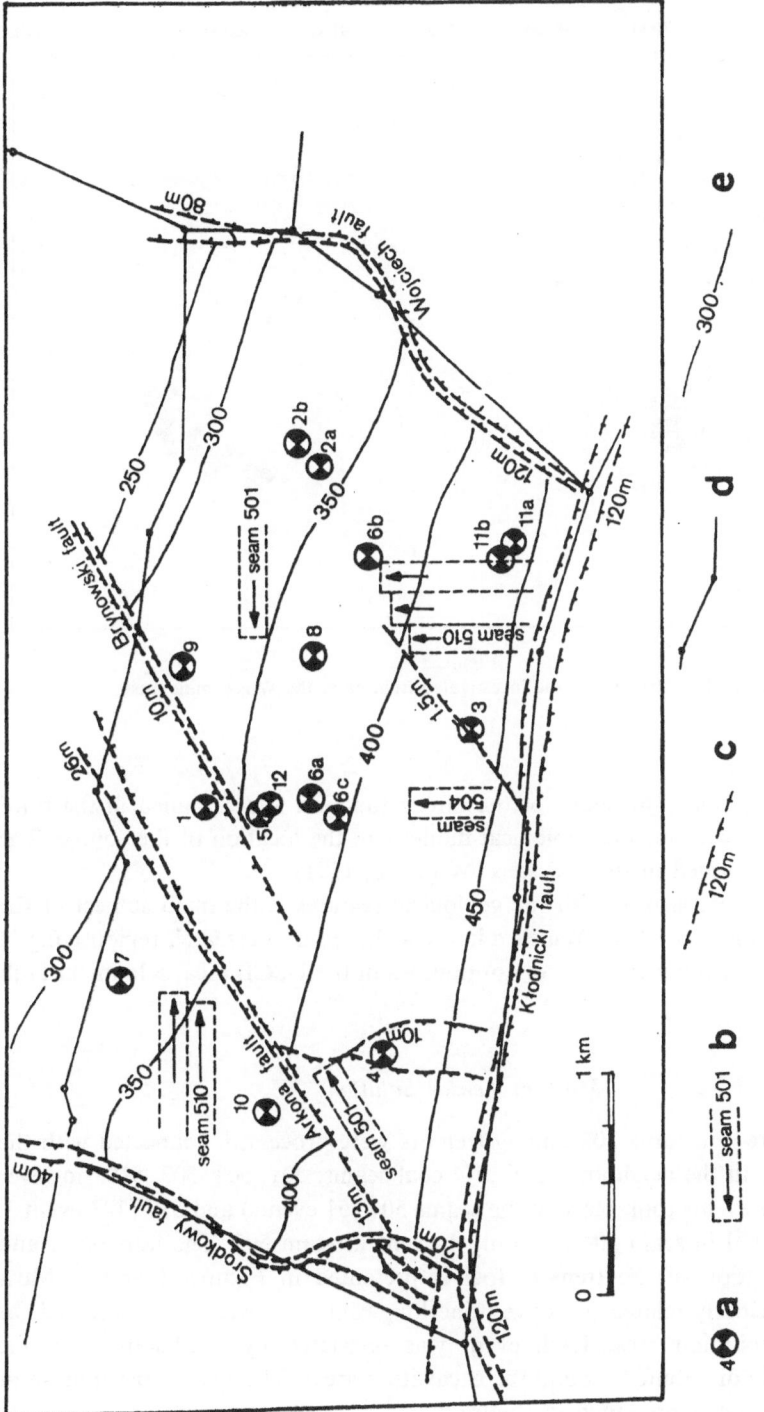

Figure 2

Geological structure of the Wujek mine area. a. Seismic stations (numbers of stations moved during the observation period were denoted as e.g. 2a, 2b, etc); b. longwall mining area; c. major faults with mean amplitude value; d. boundary of the mine area; e. isarithms delineating configuration of the roof of seam 510 with contour intervals of 50 m.

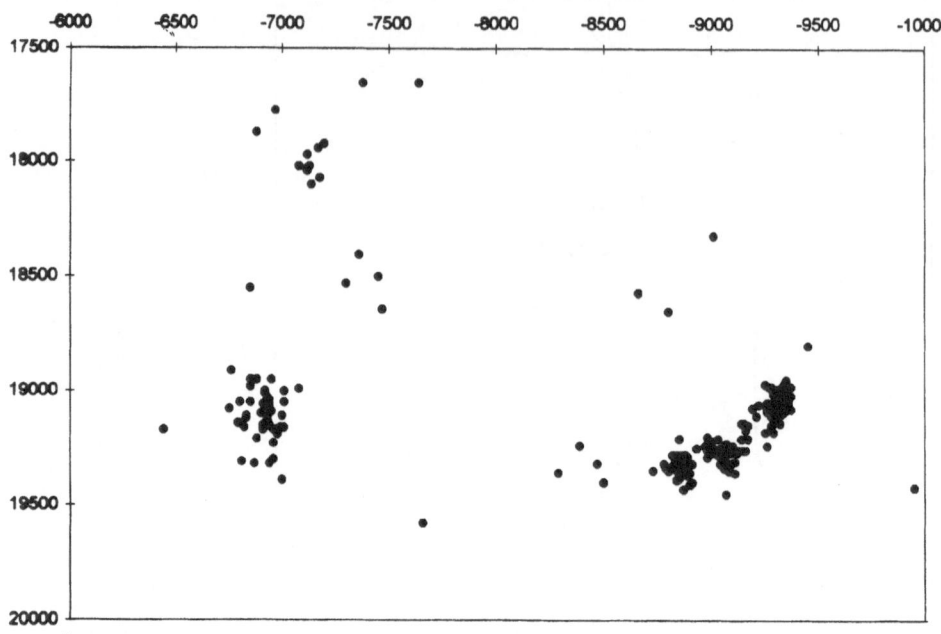

Figure 3
Horizontal localization of the investigated tremors in the Wujek mine area.

the particular station. The shape of the Green function is determined in the third step, taking into account the geological model and the location of the source. The inversion is performed in the last step (WIEJACZ, 1991).

Because the comparison with the geological features is the main subject of the present paper, the area of the Wujek mine was chosen as a research region lying in the zone of one of the main fault discontinuities in the USCB area, Klodnicki fault (Figs. 1 and 2).

Moment Tensor Solutions

In the entire mine area 202 mining tremors were processed, connected with the mining activity in the neighbourhood of 4 coal seams: 501, 504, 507, 510, however the majority of events took place in the seams 501 (51 events) and 510 (143 events). 7 events occurred in seam 504 and only 1 event in seam 507. The horizontal and vertical localization of the tremors foci is presented in Figures 3 and 4. Most tremors were closely related to active mine longwalls, however a few occurred far from the exploitation area. Each event was registered by a 12-station seismic network, evenly distributed around the excavation areas. The spatial distribution of the seismic station is presented in Figure 2.

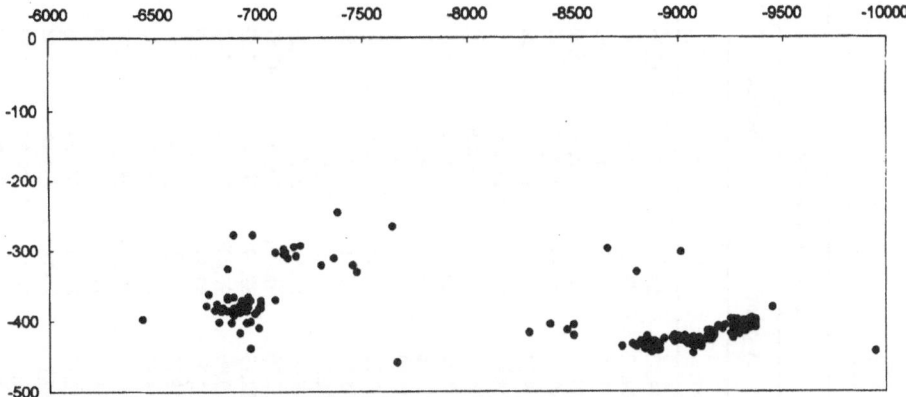

Figure 4
Vertical localization of the investigated tremors in the Wujek mine area.

Digital seismograms were used to calculate the shape of the seismic moment tensor, as well as the proportion of the three different components causing the fracture process at the focus: the isotropic component (I) corresponding to the volumetric change; the compensated linear vector dipole (CLVD) corresponding to

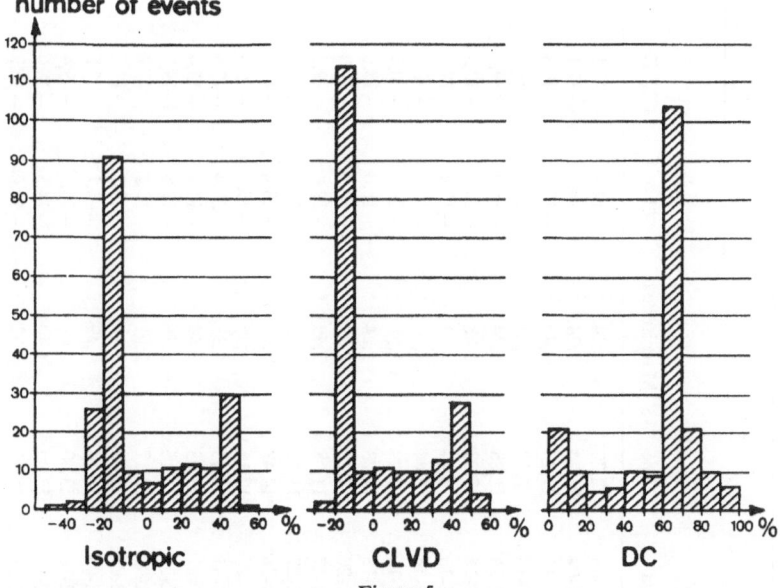

Figure 5
Shares of the particular components of the seismic moment tensor for the 202 mining tremors from the Wujek mine area for full moment-tensor solution: from the left: isotropic part; compensated linear vector dipole part; double-couple part.

Table 1

The obtained mechanism of the mining tremors from the Wujek mine area

Data	Time	Magnitude	Full Tensor			Zero trace		Double-couple			
								A plane		B-plane	
			1	CLVD	DC	CLVD	DC	φ	δ	φ	δ
93-01-03	12,22	1,24	−14,4	−17,1	68,5	−10,2	89,8	214,02	88,93	307,34	17,95
93-01-04	21,24	1,24	−14,4	−17,1	68,5	−10,2	89,8	214,02	88,93	307,34	17,95
93-01-11	21,29	1,40	−19,4	−19,3	61,3	−8,6	91,4	343,58	88,47	221,16	2,85
93-01-13	6,39	1,24	−17,8	−19,7	62,5	−7,2	92,8	340,90	83,89	132,14	6,96
93-01-15	11,28	1,24	−18,0	−10,4	71,6	−17,7	82,3	17,32	86,66	239,07	4,47
93-01-16	3,03	1,68	−23,9	−12,3	63,8	31,7	68,3	9,34	80,20	194,74	9,84
93-01-16	5,44	1,33	−19,9	−19,9	60,3	−5,9	94,1	338,97	85,24	135,78	5,18
93-01-16	6,44	1,76	−17,2	−18,3	64,5	−1,5	98,5	348,12	88,42	233,03	3,72
93-01-16	7,29	1,55	−19,2	−18,4	62,5	−16,9	83,1	2,56	86,96	218,10	3,74
93-01-18	11,23	1,24	−18,0	−18,6	63,4	−5,2	94,8	343,68	88,18	224,87	3,77
93-01-20	17,30	2,29	−18,7	−19,0	62,3	3,7	96,3	341,29	88,28	226,58	4,10
93-01-27	11,08	1,49	−19,4	−19,7	60,9	−3,8	96,2	347,20	89,96	0,00	0,04
93-01-27	11,15	1,49	23,8	21,6	54,6	−6,4	93,6	348,40	88,51	233,97	3,60
93-01-28	14,48	1,49	−18,4	−17,6	64,0	3,7	96,3	332,26	88,26	236,14	15,90
93-01-28	17,16	1,49	−19,4	−19,5	61,1	−5,8	94,2	247,06	89,51	250,53	4,29
93-01-31	10,55	1,33	−21,4	−17,6	61,0	64,9	35,1	135,76	51,74	350,64	43,87
93-01-31	23,05	1,40	−11,2	−15,4	73,4	−4,9	95,1	12,13	89,09	261,09	2,53
93-02-02	6,16	1,76	−19,8	−19,7	60,4	−2,8	97,2	0,00	90,00	269,94	2,03
93-02-02	23,17	2,01	12,2	11,2	76,6	−2,1	97,9	182,74	83,68	66,72	14,17
93-02-03	5,40	1,45	−19,8	−19,7	60,5	0,6	99,4	355,76	89,15	251,56	3,46
93-02-08	16,14	1,49	−8,2	0,6	91,1	11,8	88,2	346,61	88,78	241,08	4,54
93-02-10	13,40	1,49	−19,6	−19,4	60,9	3,7	96,3	346,72	89,59	250,48	3,77
93-02-11	13,05	1,40	−14,5	−18,0	67,4	−5,7	94,3	155,11	89,02	329,11	0,98
93-02-11	18,11	1,33	−12,0	−15,2	72,7	−4,4	95,6	352,42	88,77	247,08	4,63
93-02-16	5,44	1,40	31,4	38,3	30,3	5,8	94,2	347,33	89,75	252,18	2,74
93-02-16	12,50	1,40	−17,7	−17,8	64,6	13,6	86,4	333,13	83,69	129,77	6,87
93-02-16	21,46	2,01	−19,1	−19,9	61,0	1,7	98,3	175,97	86,81	64,29	8,57

Date											
93-02-18	6,09	1,33	21,4	37,1	41,5	15,6	84,4	158,33	89,43	265,32	1,96
93-02-18	18,06	1,40	-19,7	-19,0	61,3	-1,3	98,7	164,69	89,49	268,04	2,21
93-02-18	19,34	1,24	30,8	42,1	27,1	65,8	34,2	56,95	81,21	312,74	32,21
93-02-19	21,57	1,40	-15,8	-13,1	71,1	-19,7	80,3	306,88	87,65	212,96	30,95
93-02-22	9,10	1,74	-20,4	-17,6	62,0	-2,3	97,7	162,78	88,88	356,43	1,15
93-02-24	7,29	1,33	-18,6	-19,4	61,9	6,7	93,3	339,57	89,01	234,46	3,78
93-02-24	14,26	1,49	-17,7	-19,8	62,4	-0,4	99,6	339,07	84,96	236,88	22,65
93-02-24	18,46	1,55	-19,0	-19,6	61,4	4,2	95,8	339,58	89,19	235,55	3,34
93-02-24	19,11	1,40	-15,1	-17,3	67,6	-3,6	96,4	158,65	90,00	248,67	4,66
93-02-25	18,17	1,24	-18,7	-19,3	62,0	7,0	93,0	346,19	89,34	248,78	5,08
93-02-26	6,24	1,24	-16,9	-14,0	69,1	11,7	88,3	0,00	90,00	269,95	2,30
93-02-26	12,51	1,40	-18,6	-19,8	61,6	2,1	97,9	347,38	89,44	249,67	4,15
93-02-26	13,00	1,40	7,3	9,7	83,1	0,5	99,5	349,67	89,14	249,97	5,10
93-03-01	16,43	1,24	37,7	38,1	24,2	14,8	85,2	152,13	87,98	43,98	6,46
93-03-02	10,25	1,24	11,5	35,0	53,5	26,7	73,3	170,79	88,47	0,00	1,55
93-03-03	10,44	1,12	-17,9	-12,8	69,4	-4,9	95,1	67,66	68,60	250,50	21,42
93-03-04	7,18	1,21	-19,3	-19,5	61,2	20,2	79,8	339,54	87,30	231,80	8,80
93-03-06	6,31	0,96	-20,0	-19,6	60,5	29,3	70,7	173,31	86,89	281,96	9,65
93-03-09	20,30	0,87	-20,3	-19,6	60,1	36,0	64,0	135,15	85,66	233,55	27,46
93-03-16	9,46	1,21	-12,8	-15,6	71,6	3,3	96,7	130,74	89,59	33,43	3,25
93-03-16	12,08	1,24	-5,0	-16,9	78,1	-17,7	82,3	245,55	76,60	119,28	21,93
93-03-17	4,37	1,87	5,1	2,33	92,5	-1,9	98,1	164,83	88,40	274,81	4,66
93-03-17	5,40	1,40	37,9	39,3	22,8	33,7	66,3	177,06	70,60	320,59	23,64
93-03-18	3,06	1,52	-19,0	-19,0	61,9	-6,7	93,3	342,35	90,00	0,00	0,02
93-03-19	5,28	1,21	-19,9	-20,0	60,1	1,0	99,0	36,41	88,90	141,53	4,20
93-03-20	3,07	1,16	-13,3	-7,9	78,8	5,6	94,4	188,86	88,04	91,40	14,81
93-03-20	4,51	1,19	-18,7	-20,1	61,2	-2,5	97,5	12,23	86,79	243,98	5,18
93-03-20	5,16	1,24	-19,6	-20,0	60,4	-1,6	98,4	41,49	89,45	138,04	4,79
93-03-22	10,53	0,87	-20,1	-19,7	60,2	3,1	96,9	190,00	89,95	0,00	0,07
93-03-22	19,13	1,08	-20,9	-19,3	59,8	9,8	90,2	22,03	88,53	153,77	2,87
93-03-23	19,14	1,24	-20,2	-19,1	60,6	-4,3	95,7	342,72	87,53	219,46	4,50
93-03-24	5,01	1,12	-19,8	-19,7	60,5	18,0	82,0	358,29	88,96	238,99	2,12
93-03-24	5,35	0,71	-20,2	-19,6	60,2	-5,1	94,9	31,88	86,90	135,61	12,84
93-03-26	4,55	1,55	-14,8	-14,5	70,7	1,9	98,1	183,24	89,98	0,00	0,02
93-03-27	0,04	1,71	-19,5	-19,2	61,3	6,7	93,3	33,04	83,58	181,60	7,52

Table 1—Continued

Data	Time	Magnitude	Full Tensor			Zero trace		Double-couple			
								A plane		B-plane	
			1	CLVD	DC	CLVD	DC	φ	δ	φ	δ
93-03-30	5,36	1,33	−20,0	−19,7	60,3	2,6	97,4	0,08	89,21	237,13	1,45
93-03-31	4,33	1,03	−20,0	−19,9	60,1	−0,1	99,9	29,07	88,32	158,12	3,47
93-03-31	5,24	1,19	−19,4	−19,3	61,3	−0,4	99,6	180,00	88,59	2,68	1,41
93-03-31	22,48	1,71	−20,4	−19,6	60,0	−0,7	99,3	31,60	87,51	139,91	7,87
93-04-13	13,10	1,49	−18,5	−18,3	63,2	2,9	97,1	181,71	89,95	0,00	0,05
93-04-17	4,30	1,97	−20,8	−18,2	61,0	−9,3	90,7	170,09	85,23	346,36	4,78
93-04-21	13,35	1,40	−19,8	−19,8	60,3	−0,5	99,5	187,25	89,94	0,00	0,06
93-04-25	1,30	1,40	−17,6	−18,0	64,4	3,0	97,0	206,85	88,62	100,54	4,90
93-04-26	20,41	1,86	−20,2	−19,7	60,1	6,5	93,5	207,59	89,10	105,08	4,16
93-04-28	8,05	1,40	−19,4	−19,9	60,6	−3,4	96,6	182,91	89,94	0,00	0,06
93-05-04	4,59	1,49	−19,8	−19,8	60,5	8,1	91,9	181,90	89,93	0,00	0,07
93-05-04	12,27	1,19	−19,9	−19,7	60,5	0,0	100,0	180,39	87,96	7,05	2,05
93-05-04	18,44	1,16	−20,0	−19,5	60,5	−0,3	99,7	180,78	88,62	359,32	1,38
93-05-04	18,58	1,24	−19,2	−18,1	62,7	5,6	94,4	187,95	89,98	0,00	0,02
93-05-05	14,12	2,45	−18,1	−18,9	63,1	1,9	98,1	0,44	89,73	264,18	2,46
93-05-10	18,45	1,49	−20,1	−19,8	60,1	3,5	96,5	197,35	89,20	315,65	1,69
93-05-12	5,59	1,33	−20,1	−19,7	60,2	7,2	92,8	187,80	89,60	0,00	0,42
93-05-12	6,03	1,16	−19,9	−20,0	60,1	2,3	97,7	215,15	89,10	60,47	1,00
93-05-12	13,12	1,24	−19,8	−19,2	61,0	−2,6	97,4	169,71	88,61	1,61	1,42
93-05-13	6,57	1,60	−19,5	−19,2	61,3	−1,4	98,6	166,46	89,95	0,00	0,06
93-05-14	5,16	1,08	−19,5	−19,2	61,3	1,2	98,8	168,23	89,64	0,00	0,47
93-05-17	5,28	1,21	−19,6	−19,1	61,3	−1,6	98,4	276,07	89,95	0,00	0,06
93-05-19	6,10	1,33	−20,0	−19,8	60,1	11,3	88,7	190,09	89,55	0,00	0,48
93-05-19	7,13	1,19	−19,7	−20,0	60,4	15,1	84,9	354,96	86,86	229,68	5,42
93-05-19	11,43	1,24	−18,6	−19,7	61,7	0,7	99,3	348,78	88,26	223,79	3,03
93-05-21	5,59	1,24	−16,4	−16,3	67,2	1,1	98,9	166,47	89,85	0,00	0,16
93-05-21	7,10	1,21	−19,9	−19,4	60,7	1,8	98,2	350,74	89,54	223,16	0,75
93-05-21	14,17	1,21	−20,0	−19,8	60,2	9,7	90,3	358,29	88,23	234,32	3,17

93-05-25	16,22	1,55	−20,4	−19,6	59,9	10,6	89,4	199,58	85,54	30,68	4,55
93-05-25	19,39	2,24	−19,2	−18,6	62,2	−1,1	98,9	158,85	89,78	0,00	0,37
93-05-26	14,39	1,65	−20,0	−19,4	60,6	5,3	94,7	173,07	89,63	0,00	0,38
93-05-26	14,41	1,16	43,3	43,4	13,3	23,9	76,1	294,99	89,03	39,55	3,84
93-05-27	8,35	1,19	−20,5	−19,7	59,9	8,0	92,0	28,66	88,75	137,15	3,95
93-05-31	14,54	1,24	−0,8	−0,5	98,7	0,3	99,7	169,54	87,14	0,00	2,91
93-05-31	20,07	1,19	−18,4	−19,3	62,3	3,1	96,6	168,52	89,96	0,00	0,04
93-06-01	10,58	1,55	−20,5	−19,6	59,9	4,2	95,8	34,39	89,55	129,23	5,26
93-06-01	11,08	1,49	−20,1	−19,9	60,0	4,4	95,6	27,42	88,54	136,03	4,58
93-06-03	12,07	2,45	10,8	39,7	49,5	31,8	68,2	288,24	88,26	39,37	4,83
93-06-04	11,23	1,65	10,3	7,0	82,6	1,8	98,2	139,71	88,11	337,96	1,99
93-06-08	12,19	1,65	37,1	16,7	46,2	17,4	82,6	139,71	88,11	337,96	1,99
93-06-14	16,47	1,71	21,7	16,6	61,6	−3,3	96,7	44,01	83,76	277,18	10,34
93-06-15	3,48	1,71	−3,6	3,3	93,1	−3,4	96,6	269,42	75,81	43,55	19,96
93-06-23	3,50	1,76	12,6	18,5	68,9	18,0	82,0	244,53	72,56	45,32	18,40
93-06-24	20,20	1,24	−22,6	−16,7	60,7	−2,7	97,3	342,86	81,03	99,22	19,56
93-06-25	0,53	1,40	0,5	0,0	99,4	−2,4	97,6	318,46	89,48	221,29	4,14
93-06-25	7,42	1,24	−20,7	−16,0	63,2	26,2	73,8	107,37	80,20	220,36	23,85
93-06-29	23,56	1,52	−19,6	−18,6	61,8	−11,5	88,5	214,58	82,71	73,76	9,37
93-07-01	6,37	1,76	−22,1	−17,5	60,3	−8,0	92,0	228,36	86,48	5,55	4,79
93-07-02	7,08	1,40	−17,7	−16,6	65,8	0,2	99,8	230,34	88,03	348,73	4,14
93-07-02	18,03	1,92	−26,4	−14,7	58,9	−10,1	89,9	222,71	88,91	326,77	44,48
93-07-07	11,30	1,24	−6,1	−9,6	84,2	0,6	99,4	318,74	89,07	215,43	4,01
93-07-07	17,56	1,55	−18,6	−14,4	66,9	−0,2	99,8	226,16	88,48	336,69	4,33
93-07-09	13,19	1,68	−20,1	−18,9	61,0	22,0	78,0	99,26	85,30	207,86	14,45
93-07-09	17,16	1,76	27,8	29,2	42,9	23,6	76,4	94,50	82,18	213,94	15,61
93-07-15	17,43	1,55	0,0	−9,0	91,0	24,1	75,9	232,70	89,42	326,61	8,46
93-07-16	10,16	1,60	−20,7	−19,3	60,0	−1,4	98,6	243,96	86,00	16,95	5,86
93-07-19	20,07	1,40	−11,4	−11,0	77,6	−0,2	99,8	234,06	88,03	342,89	6,07
93-07-19	21,32	1,55	−6,0	40,3	53,7	46,2	53,8	101,72	78,71	206,81	37,49
93-07-21	20,21	1,76	−17,1	−20,4	62,6	−0,2	99,8	57,76	89,41	320,03	4,35
93-07-23	20,20	2,45	−20,8	−19,4	59,9	12,6	87,4	277,21	85,78	11,32	45,88
93-07-25	4,00	2,24	−17,2	−16,8	66,0	0,7	99,3	223,20	87,04	346,04	5,44
93-09-09	11,53	1,71	−9,6	−15,9	74,5	−1,0	99,0	159,83	86,39	314,70	3,99
93-09-10	18,37	1,92	−5,6	−9,4	85,0	−6,8	93,2	199,49	86,95	299,74	16,66
93-09-11	11,10	3,29	47,5	47,5	5,0	−7,2	92,8	81,44	88,52	178,72	11,52

G. Sagan *et al.*

Table 1—Continued

Data	Time	Magnitude	Full Tensor			Zero trace		Double-couple			
								A plane		B-plane	
			1	CLVD	DC	CLVD	DC	ϕ	δ	ϕ	δ
93-09-13	8,24	1,26	-18,1	-17,5	64,4	-22,4	77,6	256,90	76,68	19,00	20,64
93-09-19	3,26	1,70	-30,5	3,0	66,6	-2,9	97,1	295,30	89,38	25,90	45,94
93-09-21	12,15	1,86	-12,5	24,3	63,2	-7,0	93,0	100,81	85,69	197,57	32,66
93-10-14	7,30	1,49	42,0	40,4	17,5	-6,3	93,7	341,56	87,71	75,16	32,51
93-10-14	12,59	1,49	45,3	51,9	2,8	0,1	99,9	248,34	85,71	90,46	4,62
93-10-15	5,17	1,03	-13,9	-14,8	71,3	1,2	98,8	303,00	87,75	198,22	8,74
93-10-19	10,15	1,60	-11,1	-12,9	76,0	7,4	92,6	26,25	88,94	259,45	1,77
93-10-19	15,45	1,60	49,7	50,1	0,2	1,2	98,8	266,47	87,85	116,29	2,48
93-10-20	9,23	1,40	-19,5	-19,3	61,2	-2,2	97,8	307,34	86,73	72,52	5,66
93-10-21	12,52	1,65	-11,2	-13,6	75,2	-6,3	93,7	174,16	89,34	90,05	9,18
93-10-21	15,11	1,49	26,6	25,3	48,1	-2,1	97,9	156,65	86,54	28,93	5,64
93-11-03	9,31	1,65	49,3	48,9	1,9	-5,0	95,0	179,00	85,20	38,39	6,20
93-11-04	9,05	1,49	-19,3	-19,9	60,7	0,5	99,5	304,74	86,62	150,51	3,75
93-11-04	14,51	1,19	46,0	45,4	8,6	-19,0	81,0	192,21	84,97	60,83	7,58
93-11-04	18,42	1,49	39,1	40,8	20,1	-8,7	91,3	50,76	65,22	255,44	26,93
93-11-05	1,32	1,71	49,9	47,5	2,6	2,3	97,7	188,14	78,33	0,00	11,79
93-11-05	3,32	1,65	50,1	48,4	1,5	-2,5	97,5	197,78	82,46	43,24	8,34
93-11-05	16,43	1,49	25,1	12,9	61,9	-12,3	87,7	231,44	84,18	66,03	6,01
93-11-06	5,52	1,65	38,4	28,6	33,0	-16,5	83,5	198,34	86,70	91,75	11,43
93-12-03	19,51	1,60	-14,9	-15,2	69,9	27,5	72,5	113,23	58,66	341,00	42,18
93-12-04	4,54	0,71	48,0	48,1	3,9	-10,5	89,5	318,39	82,94	198,03	13,77
93-12-04	20,02	0,96	4,7	10,4	85,0	4,5	95,5	179,35	89,38	69,92	1,87
93-12-05	5,36	2,01	28,0	28,6	43,5	-1,1	98,9	299,26	87,69	195,44	9,58
93-12-06	18,58	1,24	33,3	36,0	30,8	7,6	92,4	350,74	87,15	126,90	3,96
93-12-07	3,05	1,40	15,9	12,6	71,5	-3,2	96,8	225,45	89,24	91,75	1,11
93-12-07	17,23	1,16	22,0	28,4	49,5	-7,9	92,1	339,12	85,68	95,87	9,52
93-12-07	18,59	1,49	43,9	43,6	12,5	1,5	98,5	346,48	89,90	78,28	3,27
93-12-08	8,59	1,03	20,3	13,0	66,7	-5,4.	94,6	170,21	82,30	69,00	34,82

93-12-09	3,14	1,55	−13,5	−13,0	73,6	2,0	98,0	186,01	89,37	85,87	3,55
93-12-09	11,01	1,12	44,7	38,7	16,6	−10,4	89,6	315,55	85,72	197,60	9,08
93-12-09	11,39	1,03	−17,8	−19,1	63,1	−6,5	93,5	52,82	87,24	190,69	3,72
93-12-09	12,39	1,08	47,4	37,0	15,6	12,2	87,8	30,40	87,38	131,63	13,24
93-12-09	16,15	1,12	4,7	−18,0	77,3	−20,1	79,9	276,35	89,00	183,36	18,46
93-12-09	19,58	1,08	−48,9	7,5	43,6	2,5	97,5	357,82	73,77	228,62	24,73
93-12-10	12,35	0,96	−19,7	−19,7	60,6	12,4	87,6	29,14	86,48	144,34	8,23
93-12-10	15,28	1,45	47,1	50,2	2,7	10,9	89,1	5,80	77,82	158,83	13,61
93-12-11	3,23	1,86	49,0	46,9	4,0	−3,5	96,5	68,51	84,77	174,20	18,70
93-12-11	17,19	1,16	47,1	44,8	8,1	−12,7	87,3	40,62	86,12	149,34	11,92
93-12-11	17,31	1,45	−4,9	−6,3	88,8	0,4	99,6	0,14	88,63	97,81	10,15
93-12-11	17,32	1,03	15,8	13,6	70,5	−1,1	98,9	270,05	89,54	175,73	6,03
93-12-11	18,38	1,12	45,6	44,6	9,8	−1,7	98,3	28,18	88,96	192,96	1,08
93-12-12	3,52	1,60	11,4	−0,9	87,7	−10,9	89,1	11,33	88,81	124,57	3,02
93-12-12	5,23	0,96	−15,2	−10,0	74,8	11,7	88,3	286,28	88,11	163,84	3,53
93-12-12	10,42	1,03	−19,5	−19,7	60,8	−4,7	95,3	296,31	86,39	147,00	4,20
93-12-13	16,23	1,24	43,0	43,6	13,5	8,5	91,5	232,58	89,02	32,82	1,05
93-12-13	21,15	0,71	46,9	47,3	5,9	−2,4	97,6	7,46	88,17	109,41	8,77
93-12-14	3,56	1,40	47,3	49,5	3,2	−2,2	97,8	50,42	79,46	180,99	15,96
93-12-14	12,32	1,40	47,5	47,5	5,0	−4,9	95,1	39,22	81,40	164,78	14,57
93-12-14	14,19	2,24	48,5	45,8	5,8	1,9	98,1	57,55	86,89	310,71	10,61
93-12-14	15,38	1,08	46,0	35,4	18,6	−18,0	82,0	83,85	80,78	198,24	21,46
93-12-14	19,26	1,08	47,4	41,4	11,2	−18,1	81,9	221,08	84,73	123,90	36,40
93-12-15	6,31	1,19	−20,0	−19,9	60,1	0,2	99,8	307,78	85,38	172,17	6,45
93-12-15	17,19	1,08	47,4	49,5	3,0	−3,9	96,1	16,12	79,21	159,31	13,39
93-12-16	7,02	1,55	27,7	38,1	34,2	11,9	88,1	54,85	75,59	184,52	21,92
93-12-16	13,17	0,96	47,1	41,9	10,9	−7,1	92,9	310,74	89,07	217,69	17,00
93-12-16	19,44	1,21	45,1	49,7	5,3	−3,4	96,6	30,4	79,67	168,36	13,75
93-12-17	4,55	1,33	33,8	34,0	32,1	−0,4	99,6	321,80	89,38	180,00	0,79
93-12-17	5,39	1,45	38,7	28,1	33,2	−7,3	92,7	61,20	81,75	281,22	10,72
93-12-17	4,43	1,16	45,9	49,6	4,5	−4,3	95,7	54,53	81,66	182,19	13,50
93-12-17	12,34	1,55	44,7	40,8	14,5	7,7	92,3	29,87	80,03	144,71	22,71
93-12-18	6,10	1,24	45,4	50,9	3,8	9,3	90,7	34,09	79,89	150,37	21,94
93-12-18	10,39	1,40	37,5	41,5	21,0	14,7	85,3	10,79	79,12	155,28	13,29
93-12-18	19,27	1,19	10,2	7,1	82,7	−5,3	94,7	45,24	80,29	171,64	16,09
93-12-18	23,16	0,87	−30,2	−4,1	65,6	27,6	72,4	232,18	72,57	354,14	30,66

Table 1—Continued

Data	Time	Magnitude	Full Tensor			Zero trace		Double-couple			
								A plane		B-plane	
			1	CLVD	DC	CLVD	DC	φ	δ	φ	δ
93-12-20	0,03	1,24	8,7	4,5	86,8	−5,3	94,7	27,47	88,85	145,43	2,46
93-12-20	11,47	1,03	45,3	48,0	6,6	1,5	98,5	41,98	78,01	174,31	17,50
93-12-21	5,47	1,24	23,5	33,8	42,8	6,1	93,9	51,29	77,32	183,87	18,40
93-12-21	6,35	1,19	18,0	21,6	60,4	0,8	99,2	179,59	75,58	64,63	31,35
93-12-21	9,50	1,21	−2,0	8,2	89,9	10,4	89,6	55,01	79,82	172,06	21,54
93-12-22	6,07	1,08	−14,6	−18,6	66,8	−14,9	85,1	283,31	87,83	79,98	2,37
93-12-23	15,53	1,92	28,1	25,5	46,3	−5,3	94,7	74,76	88,84	185,36	3,29
93-12-23	18,15	1,49	−22,0	−17,0	61,1	36,7	63,3	325,11	67,24	220,66	59,26
93-12-26	23,07	1,03	15,7	17,2	67,1	2,4	97,6	90,00	87,85	203,59	5,35
93-12-26	5,47	1,40	−18,0	−18,7	63,3	−4,5	95,5	62,76	87,86	220,23	2,32
93-12-29	12,16	1,40	−14,3	−14,8	80,9	1,0	99,0	327,86	86,65	216,14	8,96
93-12-30	6,01	1,08	−20,5	−9,5	70,4	−17,0	83,0	29,87	82,47	129,34	38,78

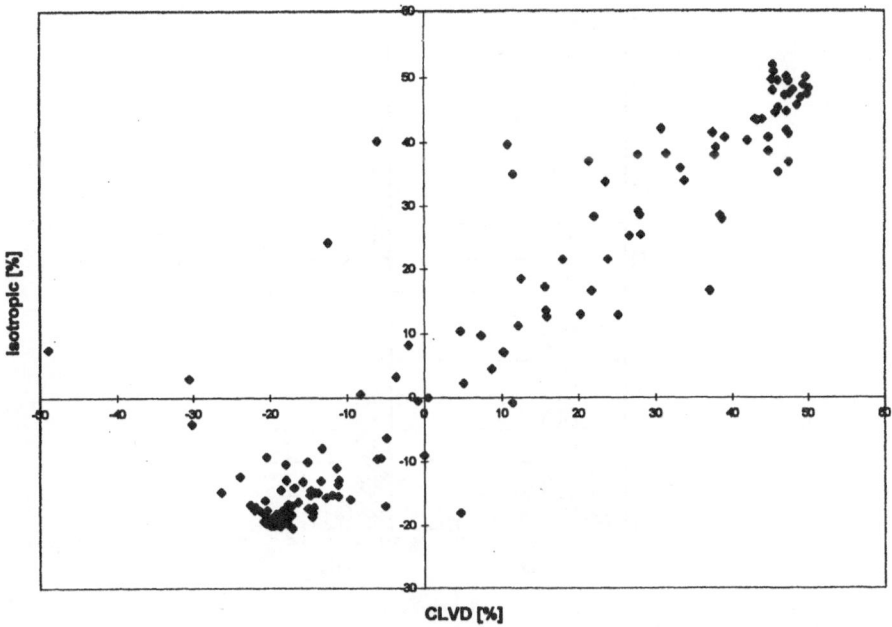

Figure 6
Relation between isotropic and CLVD components for the investigated set of mining tremors.

the uniaxial compression; and the double-couple (DC) component corresponding to the shear failure. Additionally the spatial orientation of the two nodal planes and the P and T axis were estimated. All calculations were carried out using both the L_1 and L_2 norms but, considering that there are no significant differences between the two types of results, L_1 results were used in the studies presented below. According to WIEJACZ (1991) the theoretical difference between L_1 and L_2 norms becomes significant when one or more entries is not correct and strongly differs from the expected one. The obtained differences are smaller than 5% in the part in which the particular types of components are calculated, and less than 10 degrees for the angular results of the spatial orientation of the DC component.

The stability of the method is good enough if the coverage of the focal sphere is good. This condition was approximately fulfilled for the Wujek mine area. The problem of stability was also discussed in a paper by GIBOWICZ (1993).

The relative proportions of the particular components of the moment tensor are presented in Figure 5 for the full moment tensor solutions. It is clearly visible that for about 70% of the events the double-couple component exceeds 60%, and that the maximum is between 60 and 70%. This corresponds with the low proportion of the other components (a maximum between -20 and -10% for the I and CLVD components Fig. 5). The other main types of events consist of those which have large I and CLVD components (between 40 and 50%) and a small DC part (less than 10%). This conclusion can be compared with the detailed table of the obtained

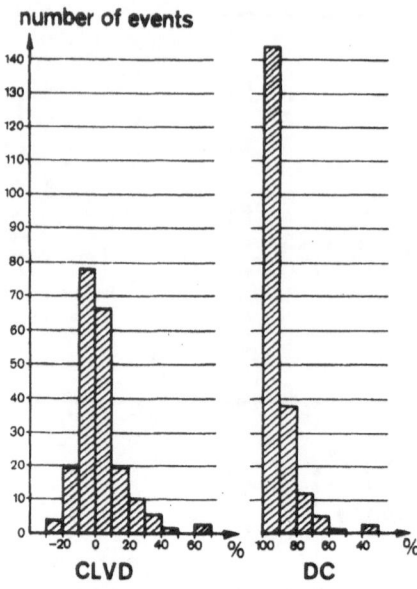

Figure 7
Shares of the CLVD and DC parts for the zero trace seismic moment-tensor solution for the 202 mining tremors from the Wujek mine area: from the left: compensated linear vector dipole part; double-couple part.

results, in which the particular types of the above-mentioned solutions can be found (Table 1). These two types of mechanisms can also be observed in Figure 6, in which points representing the I and CLVD components are concentrated in two areas. In the case of the second type (great I and CLVD components) it is possible that the obtained solutions can result from errors in the two basic assumptions of the moment tensor inversion as was considered by GIBOWICZ (1992), i.e., if the earthquake cannot be represented as a point source; or if the assumed model of rock-mass structure is incorrect; the apparent moment tensor may contain a large non-double-couple component, even if the DC component is dominant. Future investigations in this area will allow us to determine whether the second type of focal mechanism is typical for mining areas or whether it is a result of false assumptions.

The decomposition of the seismic moment tensor into the CLVD and DC components was also performed (zero trace tensor), removing any isotropic component from the solution, as well as the calculation of the tensor with both trace and determinant brought to zero (i.e., only the DC component was taken into consideration). Such a decomposition demonstrates the unimodal distribution of the dominant DC component which is presented in Figure 7.

The conclusion regarding the relatively large double-couple component has very important implications for the earlier research based on the classic fault plane

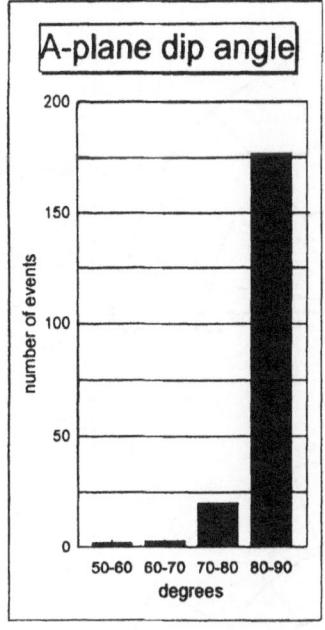
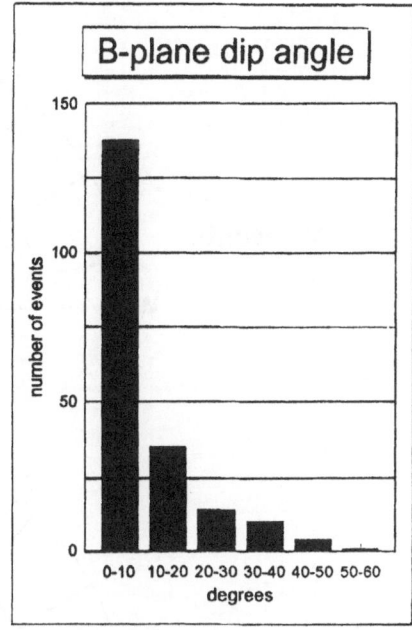

Figure 8
The angular distribution of the A- and B-nodal plane dip angle, from the left: A plane; B plane.

solution of the *P*-wave first motions (TEPER *et al.*, 1992). From the geological (or rather tectonic) point of view the shear failure is most important as a mechanism which can be caused both by mining activity and the regional stress field resulting from the lithosphere tectonic stress. The I and CLVD components seem to be typical mining mechanisms related to the collapse of wide underground openings or to pillar fracture. As we can observe, the large number of mine events with a significant DC component cannot give us the answer as to whether the tectonic stress field is the main cause of the seismic activity, until we investigate the orientation of the nodal planes obtained from the focal mechanism solution.

The spatial orientation of the two nodal planes, as well as the orientation of the two main axes, were obtained for 202 mine tremors. In this case, only the information on the DC component was used for further studies, and the results from the zero trace moment tensor solution were taken into consideration. For this reason the term "nodal plane" has been used instead of the term "nodal line," which is more appropriate for the moment tensor solution. Assuming that the dip of the A plane is always bigger than that of the B plane, the distribution of the dip angles of both planes is presented in Figure 8.

The main type of focal solution has one approximately vertical plane and one approximately horizontal plane. The selection of tremors, based on the assumption

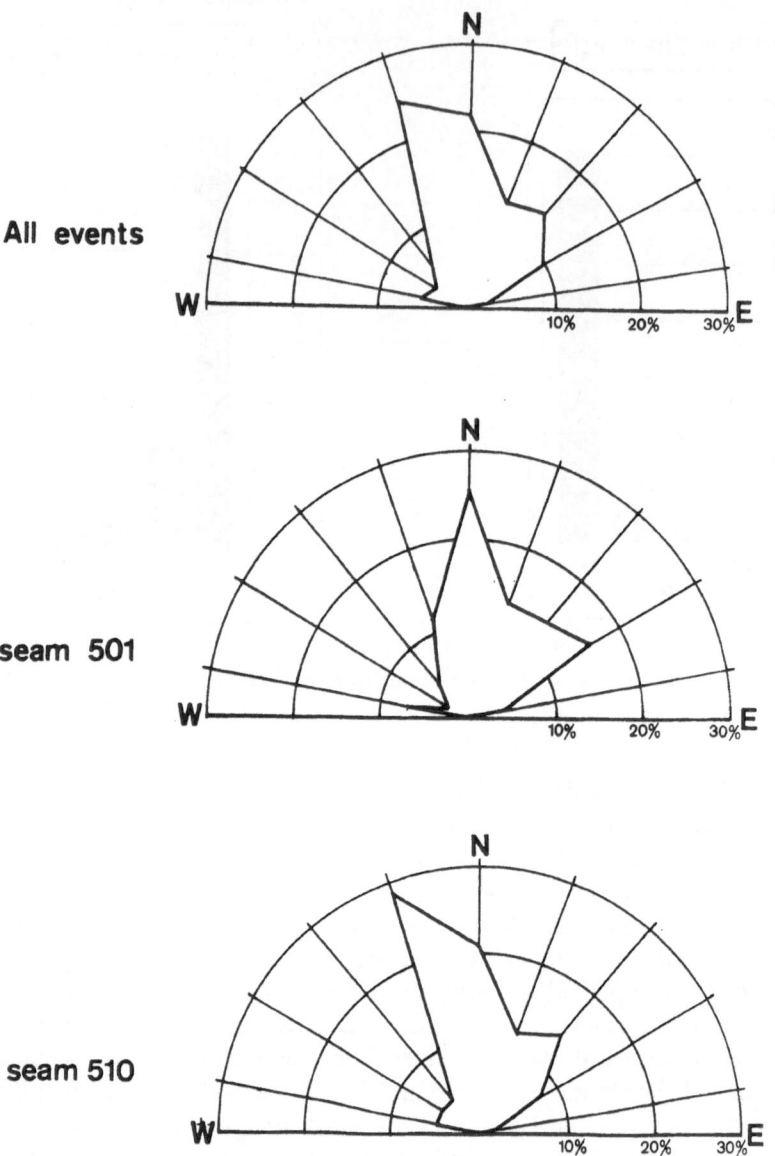

Figure 9
The distribution of the vertical plane strike azimuths for the tremors of type 1; from the top: for the entire mine area; for the events located in the vicinity of seam 501; for the events located in the vicinity of seam 510.

that the A plane has a dip angle of more than 70° and the B plane has a dip angle less than 20°, gave us 173 events. This type of mechanism is denoted as Type 1. If we assume that the vertical plane is the fault plane, the azimuth distribution of

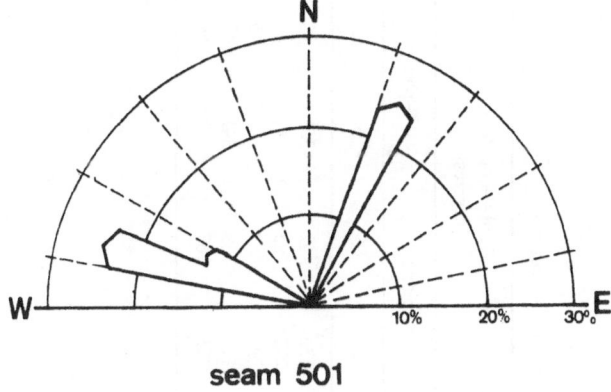

seam 501

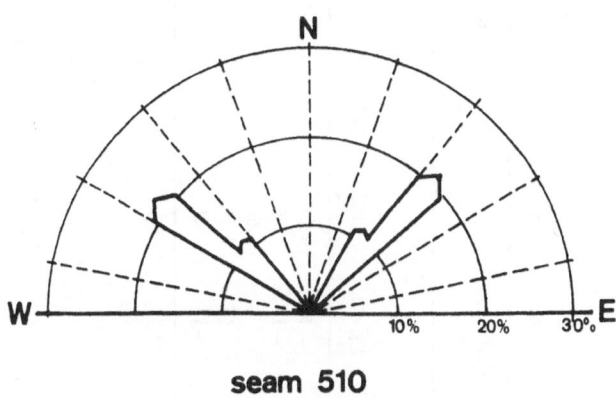

seam 510

Figure 10
The azimuthal distribution of the fracture observed in the roof of coal seams; from the top: seam 501;
seam 510.

those planes can be comprised within the geological directions and/or the directions of mining activity. The strike azimuths of the vertical planes of the shear components obtained are presented in Figure 9, where the discrimination between events located in the vicinity of seams 510 and 501 was also taken into consideration.

Two major directions are clearly dominant in the azimuth distribution: N-S (azimuth about 340–360°), and the minor one NE-SW (azimuth about 40–60°). The small clockwise rotation of those directions is observed for events from the vicinity of seam 501. The directions are not related to the trends of longwall excavation (cf., Fig. 2) whereas similar distribution of fractures orientation in the roof of seams 501 and 510 can be observed (Fig. 10).

Table 2

Results of the focal mechanism determination for 5 events with the dip of B plane greater then 40 degrees

Number (Fig. 11)	Date	Time	Mag.	Full Tensor [%]			Zero trace [%]		Double-couple [degrees]			
									A plane		B-plane	
				I	CLVD	DC	CLVD	DC	ϕ	δ	ϕ	δ
4	31 Jan 93	10,55	1,33	−21	−18	61	65	35	136	52	351	43
3	23 Jul 93	20,20	2,45	−21	−19	60	13	87	277	85	11	46
2	19 Sep 93	3,26	1,70	−31	3	66	−3	97	295	89	26	46
5	3 Dec 93	19,51	1,60	−15	−15	70	28	73	113	59	341	42
1	23 Dec 93	18,15	1,49	−22	−17	61	37	63	325	67	220	59

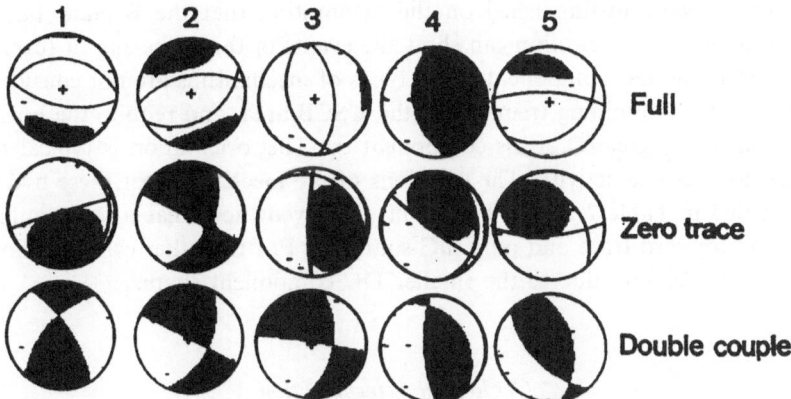

Figure 11

The seismic moment-tensor solution for selected five mine tremors from the Wujek mine area (lower hemisphere, Schmidt equal area projection). Analyzing double-couple solution, tremors nos. 1, 2, 3 indicate strike-slip mechanism, tremors nos, 4 and 5 indicate reverse, NW-SE oriented fault mechanism.

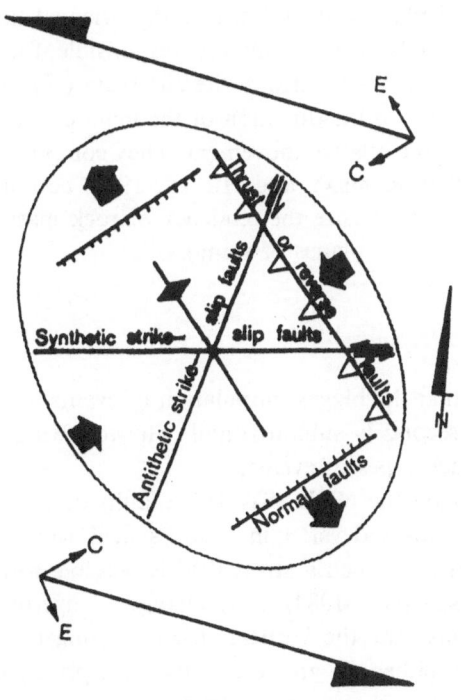

Figure 12

Pattern of neotectonic deformation of the USCB area generated by force couple acting horizontally along the fracture of crystalline basement.

Five events were distinguished on the assumption that the B plane has a dip greater than 40°. This selection can show the events of the strike-slip or reverse-slip mechanism in the DC component. Both types of mechanisms are not considered as typical ones for the mining tremors for the USCB area, and require the horizontal orientation of the greatest stress component σ_1. This orientation is typical for the regions with tectonic activity. The solutions of the mechanism for these five events are presented in Table 2 and in Figure 11. The event no. 2 has a very similar DC orientation for zero trace and pure DC solutions. For the other cases the final DC solution is less certain due to the smaller DC component share.

Geological Interpretation

Strike-slip Focal Mechanism

The solution of the last group of tremors points to a strike slip as a cause of the events. The tremors of this group were generated due to horizontal, dextral tectonic transport on NNE-SSW oriented discontinuities of rock mass and/or due to horizontal, sinistral displacement on latitudinally oriented faults. Movement along NNE-SSW oriented faults can be mainly responsible for this kind of seismic activity. NNE-SSW tending fractures in the 501 seam (Fig. 10) form "en echelon" planes lying at about 15° in the direction of the main displacement, with the acute angle pointing against the relative movement. They compose a typical R-type shear set (e.g., RIEDEL, 1929; RAMSAY and HUBER, 1984) accompanying sublongitudinal, dextral faults and emphasize the tendency of rock mass to shearing along the directions marked by nodal planes of tremors.

Normal Fault Focal Mechanism

It is no surprise that the biggest population of events has one subvertical nodal plane and that the second is subhorizontal. The following phenomena affect the largest number of such types of events.
—vertical movement on the NNE-SSW ledge faults described above.
  Typical strike-slip faults develop in a stress field with a vertical intermediate principal stress axis, σ_2. Such a stress field is possible within a narrow range of conditions (JAROSZEWSKI, 1984). Any change in overburden pressure outside certain limits means that the vertical stress no longer retains an intermediate value, but becomes either the greatest or the least principal stress. Consequently dip-slip faults are formed and not strike-slip ones.
—vertical movement on the fractures which build a set of R shears associated with NE-SW tending, normal faults.

According to basic structural models, planes of such shears are steeper than normal fault surfaces (i.e., are subvertical or vertical) and their strike follows that of the main structure.

Reverse Fault Focal Mechanism

The solution of some tremors suggests that displacement on NW-SE oriented reverse faults may occur (cf., Figure 11, cases 4 and 5). It is an evidence of occurrence of local or regional compression with a NE-SW axis direction.

Final Remarks

Neither tectonic transport directions nor relative movements listed above contradict the model of the present dynamics of the USCB postulated earlier (TEPER et al., 1992). According to the model, neotectonic deformation in the region is a typical product of force couple that has acted along the deep-rooted fracture of crystalline basement situated evenly with the latitude. The force couple has caused sinistral horizontal tectonic-transport in basement since Tertiary. As a consequence it has caused a rise in activity of various fault sets in the Carboniferous rock mass according to the pattern in Figure 12. The tectonic activity, as one of the components which cannot be ignored, results in a generation of seismic events whose types of focal mechanism, as well as spatial distribution, are well-defined and true to the model.

Acknowledgements

This research was funded by Polish Government KBN Programme (grant number KBN 9 S602 045 03).

REFERENCES

FEIGNIER, B., and YOUNG, R. P. (1992), *Moment Tensor Inversion of Induced Microseismic Events: Evidence of Nonshear Failures in the* $-4 < M < -2$ *Moment Magnitude Range*, Geophys. Res. Lett. *19*, 1503–1506.

GIBOWICZ, S. J. (1992), *Seismic Moment Tensor and its Application in Mining Seismicity Studies: A Review*, Acta Montana *88*, 37–69.

GIBOWICZ, S. J., *Seismic moment tensor and the mechanism of seismic events in mines*. In *Rockbursts and Seismicity in Mines* (Young, ed.) (Balkema, Rotterdam 1993) pp. 149–155.

GIBOWICZ, S. J., and KIJKO, A., *An Introduction to Mining Seismology* (Academic Press 1993).

JAROSZEWSKI, W., *Fault and Fold Tectonics* (PWN-Polish Sci. Publishers, Warszawa, Ellis Horwood, Chichester 1984).

McGARR, A. (1992a), *An Implosive Component in the Seismic Moment Tensor of a Mining-induced Tremor*, Geophys. Res. Lett. *19*, 1579–1582.

McGARR, A. (1992b), *Moment Tensors of Ten Witwatersrand Mine Tremors*, Pure and Appl. Geophys. *139*, 781–800.

RAMSAY, J. G., and HUBER, M. I., *The Techniques of Modern Structural Geology, v. 2* (Academic Press, London 1987).

RIEDEL, W. (1929), *Zur Mechanik geologischer Brucherscheinungen*, Zent. Min. Geol. *8*, 239–254.

SILENY, J. (1989), *The Mechanism of Small Mining Tremors from Amplitude Inversion*, Pure and Appl. Geophys. *129*, 309–324.

TEPER, L., IDZIAK, A., SAGAN, G., and ZUBEREK, W. M. (1992), *New Approach to the Studies of the Relations between Tectonics and Mining Tremors Occurrence on Example of the Upper Silesian Coal Basin (Poland)*, Acta Montana *88*, 161–177.

WIEJACZ, P. (1991), *Investigation of Focal Mechanisms of Mine Tremors by the Moment Tensor Inversion*, Ph.D. Thesis, Inst. Geophys. Pol. Acad. Sci., Warsaw, Poland (in Polish).

WIEJACZ, P. (1994), *Moment tensors for seismic events from Upper Silesian coal mines, Poland*. In *Abs. European Seismological Comm. XXIV General Assembly* (Univ. of Athens, Athens 1994) p. 99.

(Received January 5, 1995, revised October 16, 1995, accepted October 24, 1995)

PAGEOPH, Vol. 147, No. 2 (1996)

0033–4553/96/020239–09$1.50 + 0.20/0

Fractal Dimension of Faults Network in the Upper Silesian Coal Basin (Poland): Preliminary Studies

ADAM IDZIAK[1] and LESŁAW TEPER[1]

Abstract—Fractal analysis of faults network, tremor foci spatial distribution as well as the Gutenberg-Richter relationship could further explain whether the biggest seismic events are connected with recent tectonic activity. Fractality of fault systems geometry, as a first step of the analysis, was tested for a part of the USCB embodying the main structural units. The cluster analysis and the box counting methods were employed.

The calculated fractal dimension of fault network was 1.98 for the whole area yet for considered structural units it was close to 1.6. The results point to similarity of studied fault pattern to river network. Faults within selected tectonic units make separate sets which have a distinct geometry and origin. The value of 1.6 is an upper limit to the fracture geometry of rocks that can be explained on the basis of Griffith energy balance concept.

Key words: Fractal dimension, faults, cool basin, Poland.

1. Structure of the Upper Silesian Coal Basin

The Upper Silesian Coal Basin (hereafter called the USCB) is situated in southern Poland, partially in the Czech Republic (Fig. 1). Carboniferous coal-bearing formation is composed of numerous sequences of sandstones, mudstones and slates. Its thickness reaches 6000 m. In the area of about 10 thousands sq. km 70 big deep collieries operate, extracting coal measures. More than 150 coal seams are documented in 1500 m depth. Moreover, there are underground mines excavating zinc and lead ores from huge deposits located in Triassic dolomites which form cap-rock of Carboniferous in the northeastern parts of the area.

The long-drawn influence of the deep basement faults system affects the pattern of tectonics of the Silesia-Cracow region (KOTAS, 1983; TEPER *et al.*, 1986; TEPER, 1988; IDZIAK *et al.*, 1991). The vast number of the main disjunctive structures has a nature of secondary faults, following older tectonic directions and reflecting a kinematics of the USCB basement block movements. The deformational network of the USCB is comprised of sets of strike-slip, oblique-slip and dip-slip faults. Faults

---

[1] Silesian University, Faculty of Earth Sciences, ul. Bedzinska 60, 41–200 Sosnowiec, Poland.

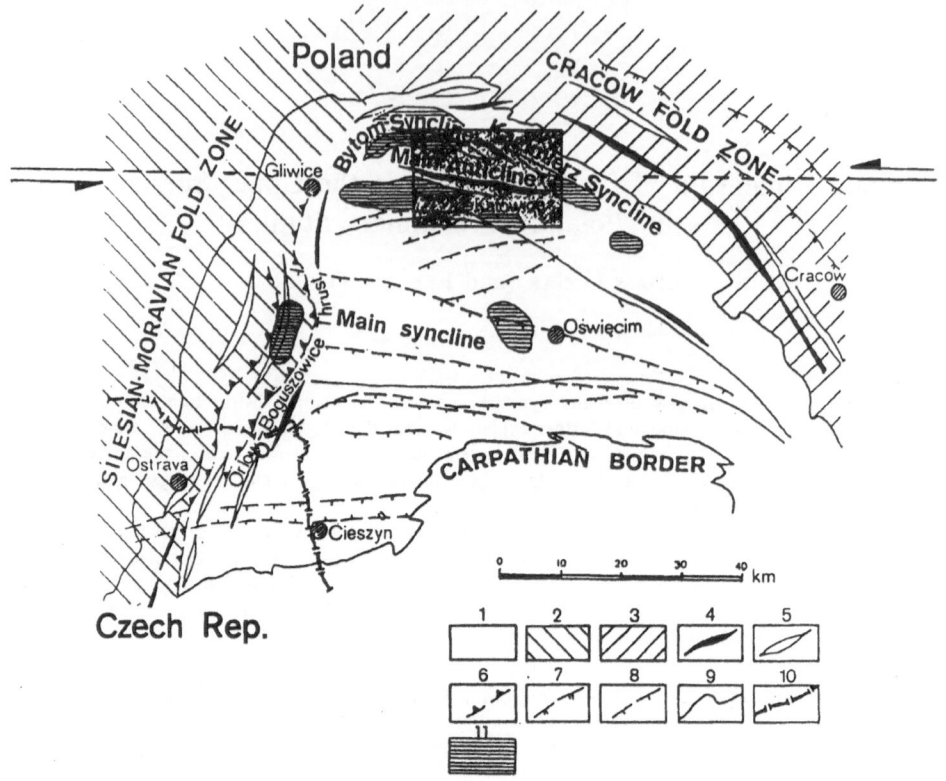

Figure 1

Structural sketch map of the USCB with location of the study area. 1, Zone of block tectonics; 2, zone of fold tectonics; 3, zone of fold-block tectonics; 4, anticlines; 5, synclines; 6, thrusts; 7, main faults; 8, main faults formed or reactivated due to Alpine movements; 9, border of the USCB; 10, state border; 11, seismic areas; the area studied (cf., Fig. 2) is shaded; arrows show directions of tectonic transport on fracture of crystalline basement.

formed during Cenozoic are a typical product of force couple that has acted evenly with the parallel of latitude since Tertiary, causing a horizontal and anti-clockwise movement (see Fig. 1) of rock-mass (TEPER *et al.*, 1992; SAGAN *et al.*, 1994).

2. *The Reason for Studying the Fractal Dimension of Fault Systems in the USCB*

The USCB is the region where great seismic activity is observed. Most of the seismic events have energy not exceeding 10^7 J. Recently they were assumed to be induced by mining activity. Their localizations and types of focal mechanism suggest the connections between the rock bursts and the generation of new discontinuities in rock complex.

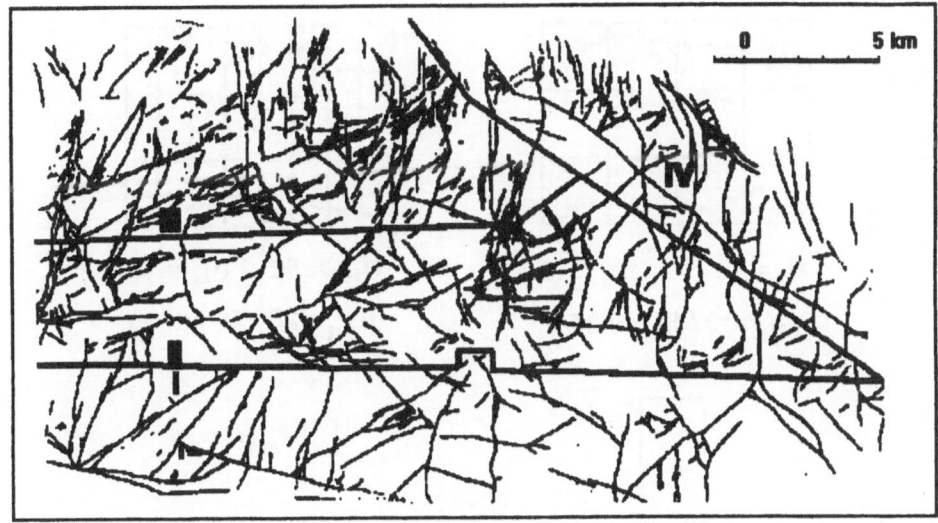

Figure 2
Digital map of fault network in the area studied (cf., Fig. 1). I, Klodnica fault system; II, Saara fault system; III. Fault sets within Bytom syncline; IV, Bedzinski fault system.

Rarely are the tremors with energy of 10^8 J or more registered. Their genesis is not yet well recognized. Some of the researchers investigating the induced seismicity of the USCB are of the opinion that the strong tremors are caused by tectonic processes which occur on a regional scale. The fractal analysis of fault systems together with the spatial and time distribution of tremors can be helpful in verification of this hypothesis.

3. The Fault Network

For a rectangular area 24 km long and 16 km wide (Fig. 1) a digital fault map 1:25 000 (Fig. 2) projected on sea level with regards to every structure (974 items) having more than 1 m of vertical throw was created. Each individual fault was mapped directly in underground mine excavations.

The area has been subdivided into four parts. The geometry of fault systems differs from one part to another as faults cut various structural units of the USCB. The distinguished parts (Fig. 2) are as follows:

1. Main anticline or system of oblique-slip Klodnica fault.
2. Transition zone or system of oblique-slip Saara fault.
3. Bytom syncline with set of NNW-SSE faults.
4. Kazimierz syncline or system of oblique-slip Bedzinski fault.

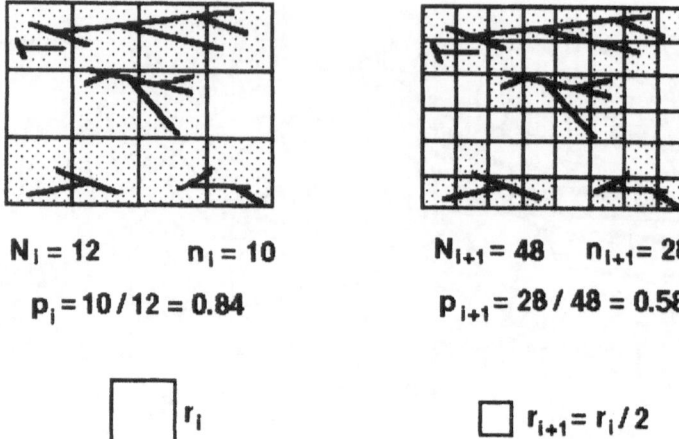

$N_i = 12$ $n_i = 10$

$p_i = 10 / 12 = 0.84$

$N_{i+1} = 48$ $n_{i+1} = 28$

$p_{i+1} = 28 / 48 = 0.58$

r_i $r_{i+1} = r_i / 2$

Figure 3

Clustering of faults, two successive steps. n, number of squares crossed by fault traces; N, total number of squares covering the area; p, 2-D geometrical probability of fault occurrence; r, box side length.

4. Clustering of Faults in the USCB

To study the clustering of faults in the USCB area a geometrical probability of fault trace occurrence was determined on the measuring plane at different scales. The fractal analysis was employed as an interpretational method. The fault map was covered by square grids with side length r_0, r_0/k, r_0/k^2, ..., (k—natural). For every grid size the number of boxes n_i traced by the fault line was counted (e.g., two steps, Fig. 3).

For any phenomenon "A" which appears on the two-dimensional surface, the geometrical probability is defined as a ratio of summarized area $S_{\text{sum}}(A)$ of parts where the phenomenon occurs on the total surface area S_{tot}:

$$p(A) = \frac{S_{\text{sum}}(A)}{S_{\text{tot}}}. \tag{1}$$

For the surface covered by squares with side length r_i the summarized area of squares traced by faults is equal:

$$S_{\text{sum}} = n_i \cdot r_i^2 \tag{2}$$

where n_i is the number of traced boxes. The total area of squares covering the surface is:

$$S_{\text{tot}} = N_i \cdot r_i^2 \tag{3}$$

where N_i is the total number of boxes. The probability that a square box of size r_i

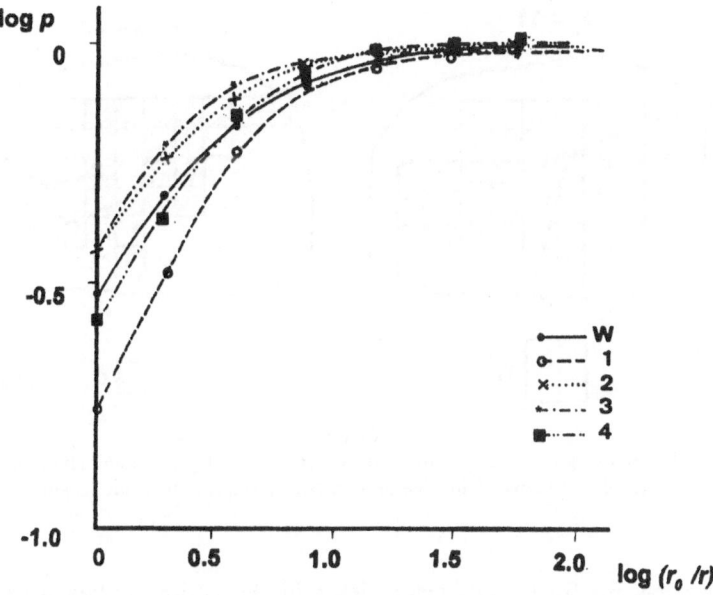

Figure 4
Log-log graphs plotting 2-D geometrical probability of fault occurrence distribution (p) versus normalized box size (r_0/r). W, for whole population; 1–4, for distinguished units (cf., Fig. 2); probability of fault occurrence has no fractal character either in the entire studied area or in any distinguished unit.

will include the trace of faults can now be estimated:

$$p_i = \frac{n_i}{N_i}. \qquad (4)$$

If the geometrical probability is self-similar it should have the fractal distribution (TURCOTTE, 1992):

$$p_i = \left(\frac{r_i}{r_0}\right)^{2-D} \qquad (5)$$

where D is a fractal dimension of probability distribution $p(r)$. The D value is a measure of fault clustering. It can change from values near zero for faults extremely concentrated in a small limited area to the value equal to 2 for faults densely spread in the entire area studied.

The fractal cluster analysis was made for the entire investigated area, as well as for the subdivided units. The analyzed areas were initially covered by boxes of 8 km side length. In every following step the box sides were two times less than in the previous step. Minimum box side length (125 m) was limited by the map scale. The obtained results show that distribution of a 2-D geometrical probability of fault

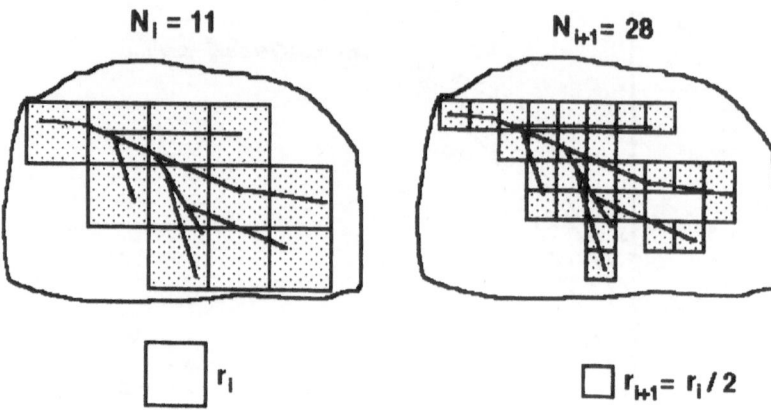

Figure 5
Fractal analysis employing box-counting method, two successive steps. N, minimum number of squares
required to cover fault line at a particular stage; r, box side length.

occurrence has no fractal character either in the entire studied area or in any
distinguished unit (Fig. 4). The division of the area studied into square boxes with
a side length exceeding 1 km yields the estimated geometrical probability equal to
1 (i.e., there is certainty of finding some fault in every box with a side longer than
1 km). For the smaller boxes, the probability sharply depends on scale effect. This
last conclusion is valid for the considered range of scales. We cannot extrapolate it
below the smallest box size considered. Thus, study of the map which scales more
than 1:25 000 should be carried out.

The box network orientation was another factor which could influence the
results. The analysis was conducted using a box grid oriented accordingly to local
Cartesian coordinates. The fault directions have an anisotropic distribution. Thus
the box network orientation might effect distribution of the 2-D geometrical
probability of fault occurrence (see e.g., VELDE *et al.*, 1990), especially for the large
box size.

5. Fractal Geometry of Fault Systems

In every geological unit the fault system geometries seem to be complex and
hard to describe. This complexity required fractal treatment. To determine the
fractal dimension of these objects, the most frequent algorithm that was applied
was that known as "box counting method" (Fig. 5).

According to the algorithm, a fault system is initially enclosed in a grid of N_0
squares with a side length r_0. In the succeeding steps the fault lines are covered with
boxes whose side length r_i decreases in sequential iterations (i.e., r_i can be equal

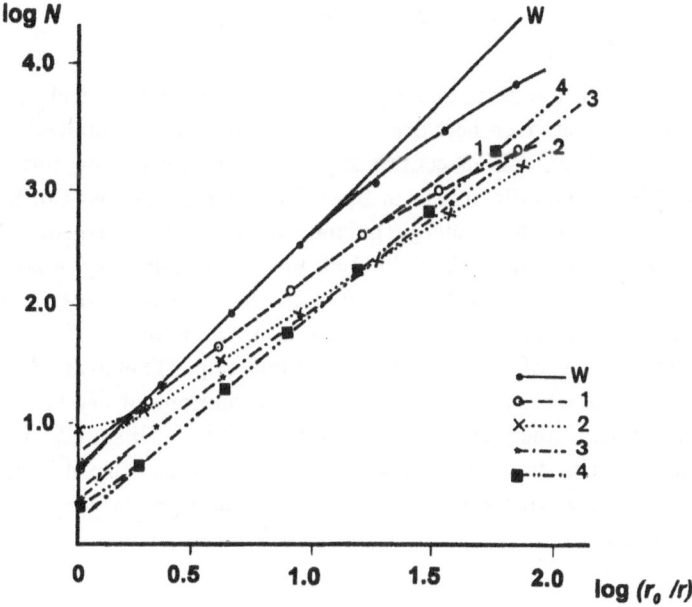

Figure 6

Log-log graphs plotting minimum number of squares required to cover fault line (N) versus normalized box size (r_0/r). W, for whole population, D value = 1.98; 1–4, for distinguished units (cf., Fig. 2), D value = 1.58, 1.58, 1.6 and 1.57, respectively.

$r_0/k, r_0/k^2, r_0/k^3, \ldots, k$ —natural). The number of boxes covering all fault lines is counted each time. If the fault system has a self-similar structure, that number should satisfy the following equation (MANDELBROT, 1983):

$$N(r_i) = N_0 \cdot \left(\frac{r_0}{r_i}\right)^D \tag{6}$$

where the exponent D is the fractal dimension of the fault system.

The $\log N - \log(r_0/r)$ graphs were plotted for the entire area studied, as well as for each of the distinguished parts of the map separately (Fig. 6). For the entire map the graph is linear in the range of box size of 1 km to 16 km (Fig. 6). The calculated fractal dimension of the fault network is 1.98. The graphs plotted for subdivided parts of the map are almost linear for box side dimension less than 8 km (Fig. 6). The fractal dimension of fault systems in considered structural units is 1.58; 1.58; 1.6; 1.57, respectively. The values of calculated fractal dimension suggest that for the entire map it is a measure of the spatial distribution of the faults, while for subdivided parts it characterizes the geometry of the fault system (branching geometry).

6. Conclusions

The fractal dimension close to 1.6 points to the similarity of each separate fault system studied, to the river network. A similar result was obtained by HIRATA (1989) for the fault systems in central Japan. According to him this value is an upper limit to the fractal dimension of the fracture geometry that can be explained on the basis of the Griffith energy balance concept. As the generation of the fracture network with a higher fractal dimension requires delivery of larger external energy to the system, D value close to 2 obtained for the entire area would evidence that external forces acted in the rock-mass more than once. This conclusion is consistent with results of tectonic interpretations (e.g., TEPER et al., 1992). The structural pattern has been controlled by relative motions of crystalline blocks of the USCB basement since Variscan time (e.g., TEPER, 1988). The relaxation of stresses could be partially performed using reactivated older discontinuities. That is probably one of the causes of the nonfractal character of the estimated 2-D geometrical probability of fault occurrence distribution.

Acknowledgements

LT thanks Christopher J. Bean and the Geological Department of University College in Dublin for their kind assistance in the map processing. The financial support of the Polish Government (Grant number KBN 9 S602 045 03) is acknowledged.

REFERENCES

HIRATA, T. (1989), *Fractal Dimension of Fault Systems in Japan: Fractal Structure in Rock Fracture Geometry at Various Scales*, Pure and Appl. Geophys. *131*, 157–170.

IDZIAK, A., TEPER, L., and CABAŁA, J. (1991), *Shallow Seismic Method as a Tool for Resolving Tectonic Problems: The Northeast Border of the Upper Silesian Coal Basin*, Acta Geophys. Pol. *39*, 293–309.

KOTAS, A., *Structural evolution of the Upper Silesian Coal Basin*. In *Proc. X Congr. Int. Strat. Carb.* C.R. *3* (Madrid 1983) pp. 459–469.

MANDELBROT, B. B., *The Fractal Geometry of Nature* (Freeman, New York 1983).

SAGAN, G., TEPER, L., and ZUBEREK, W. M., *Tectonic analysis of the mine tremor mechanisms from the Upper Silesian Coal Basin (Poland)*. In *Proc. European Seismological Commission 24th General Assembly* (Univ. of Athens, Athens 1994) pp. 1288–1289.

TEPER, L., *New results of tectonic research in NE part of the USCB*. In *Proc. IInd. Conf. Application of Geophysical Methods in the Mining Industry* (Mining Metall. Acad., Cracow 1988) pp. 291–301.

TEPER, L., HOLLEK-IDZIAK, J., and IDZIAK, A. (1986), *Joint in the Lower Coquinoid Limestone over the Kraków-Myszków Fault*, Rudy Metale 4, 106–109.

TEPER, L., IDZIAK, A., SAGAN, G., and ZUBEREK, W. M. (1992), *New Approach to the Studies of the Relations between Tectonics and Mining Tremor Occurrence on Example of the Upper Silesian Coal Basin (Poland)*, Acta Montana *88*, 161–177.

TURCOTTE, D. L., *Fractals and Chaos in Geology and Geophysics* (University Press, Cambridge 1992).
VELDE, B., DOUBOIS, J., TOUCHARD, G., and BADRI, A. (1990), *Fractal Analysis of Fractures in Rocks: The Cantor's Dust Method*, Tectonophysics *179*, 345–352.

(Received December 12, 1994, revised October 16, 1995, accepted October 30, 1995)

PAGEOPH, Vol. 147, No. 2 (1996)

0033–4553/96/020249–13$1.50 + 0.20/0

Characteristics of Mining Tremors within the Near-wave Field Zone

JÓZEF DUBIŃSKI[1] and GRZEGORZ MUTKE[1]

Abstract — Mining in the Upper Silesian Coal Basin, Poland, induces tremors with magnitudes ranging up to 4.3. The recordings of mining tremors from the near-wave field and the far-wave field show different characteristics of ground motion.

Knowledge of these characteristics, has an enormous practical importance when solving the problems of resistance of underground workings against seismic impacts. The near-wave field is characterized by a domination of the high frequency components of motion, and the seismograms often are a single pulse. Geomechanical interpretation of them leads to the conclusion that this pulse corresponds to a single exciting force. This fact is also reflected in the rockbursts: the potentially damaging ground motion which is restricted to the seismic source region.

The seismometric data have shown that the peak particle velocities from a hypocentral distance of 200–300 m, resulted in more than 3 cm/s. The peak ground velocities appear to be dependent on stress drop $\Delta\sigma$, and peak particle velocities reaching 5 m/s may occur. The computational example proves that in a thick coal seam the pulse with the stress $\sigma = 1.8$ MPa from the tremor at the short source distance can generate the rockburst.

Key words: Mining tremors, rockburst, near-wave field, peak particle velocity, dynamic load.

Introduction

Coal mining in the Upper Silesian Coal Basin induces 1000–2000 of seismic events every year with a local magnitude $M \geq 1.5$. Only 10–20 events of that number are rockbursts resulting in severe injuries of miners and extensive damage to mine workings. In the case of underground mine working stability, defined by the system supports—rock mass, the knowledge of the real characteristics and values of ground motion parameters, generated by seismic events in such a system, is the key factor in gaining an improvement and progress in this field.

At present, the largest number of rockbursts is recorded in mine roadways, which are substantially worse supported as compared to longwall workings. The analysis of the the rockburst which occurred shows that they are directly caused by the shocks localized in overlying roof strata. In the conditions of the Upper Silesian Coal Basin those are thick-beds (20–200 m) of rigid sandstones with compression strength up to 100 MPa.

[1] Central Mining Institute, Plac Gwarków 1, 40-166 Katowice, Poland.

Two main types of such rockbursts, in view of the generated effects, are mostly distinguished, namely:

- those causing destruction of the mine roadways and miners injuries (as a result of an explosive disintegration of coal initiated by the pulse wave reaching the working),
- those causing injuries and fatal accidents, with no destruction or extensive damage observed within the roadways (high ground motion from roof tremors).

The analysis of seismograms indicates that the hypocenters of the shock leading to a rockburst are localized within the roof layers to a distance of 200 m from the effects in the working, and the magnitude of those events is, as a rule, in the range of 1.5–3.0. Stress concentration zones resulting from past mining activity, are located in the working where the destruction appears most often. These are for example, the edges formed after previously conducted mining process, left pillars or natural disturbances (e.g., faults).

It follows from the above statements that only an analysis of real parameters characterizing the vibrations and transient stresses in the near-wave field, together with the knowledge of the static stress state can make progress in developing the rockburst criteria. Unfortunately, a serious lack of suitable records of the shocks from the direct vicinity of the focal zone has been observed. For this reason, several recordings which are potentially interesting for those engaged in the problem, are cited in the paper.

Examples of Recordings and Analysis of Mining Tremors within the Near-wave Field

The seismograms presented here delineate the tremors which occurred at a depth of 800 m, that is from the 21E and 22E longwall areas of the "Szombierki" mine.

One group of the records is composed of the far field recordings (500 m and more from the source of the tremor, for the event of $M > 1.5$), characterized by a good shaping of different wave groups, relatively low frequency of the dominant ground motion (2–5 Hz) and low maximum particle velocity (and consequently—the small values of the dynamic stress field). Figures 1a and 1b illustrate an example of such a record (hypocentral distance being $R = 1100$ m) for the tremor with the magnitude $M = 2.2$ ($E = 1 * 10^6$ J). The mechanism of the phenomenon has been defined as dip-slip normal with the azimuth being nearly parallel to the edge (seam 501), thus creating the cause of the formation of a stress concentration zone in the area. The parameters of this tremor (r—defined for Madariaga's model, $\Delta\sigma$— Brune's stress drop) are:

$$r = 130 \text{ m}$$

$$M_0 = 6.25 * 10^{12} \text{ Nm}$$

$$\Delta\sigma = 1.18 \text{ MPa.}$$

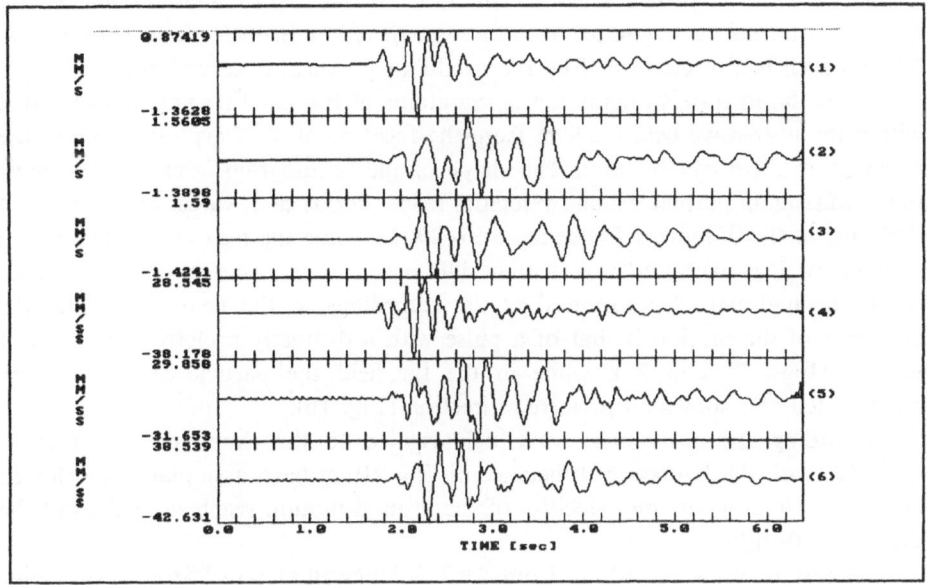

Figure 1a
Velocity (1, 2, 3) and acceleration (4, 5, 6) for a tremor of magnitude $M_L = 2.2$ (in a far field).

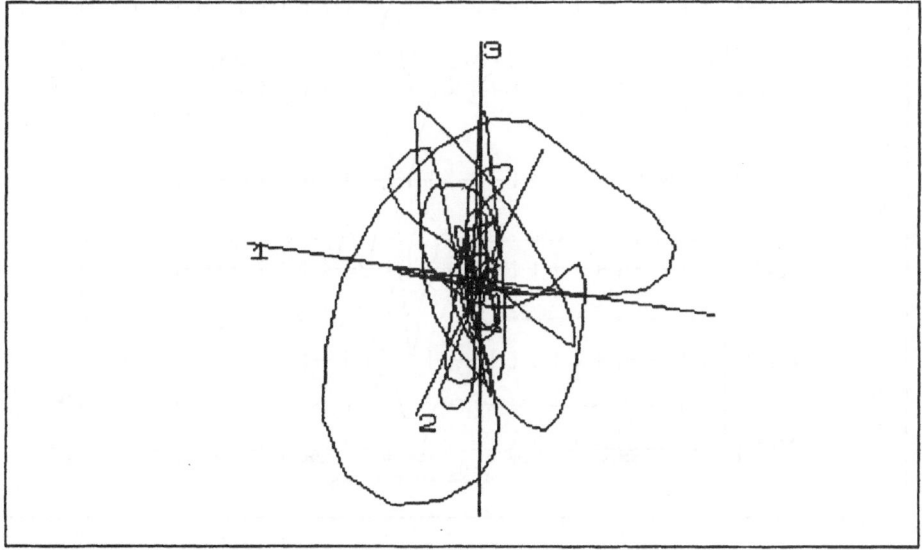

Figure 1b
Particle ground motion (velocity) for a tremor of magnitude $M_L = 2.2$ (in a far field).

Madariaga's model provides, under conditions of the Upper Silesian Coal Basin, more real dimensions as compared to those predicted by Brune's model (GIBOWICZ, 1989), which is also confirmed by the results of practical observations.

Figure 2a, in turn, presents the seismograms of the same tremor, but recorded within the near-wave field—300 m from the focal point. A sharp difference in the seismogram form can be seen. The shape of the seismogram is close to a single pulse. Maximum particle ground velocities reach, in this case, values of the order of 0.03 m/s, at the dominant frequency of 5–20 Hz, while the maximum acceleration of the particle is over 5 m/s². A distinct difference can also be seen in the character of the ground particle motion. For the recordings in the near-wave field, the character of the motion is that of a pulse with a distinctly preferred direction of motion (Figs. 2a and 2b), while in the far field the particle motion is very complicated and shows no preferred direction (Fig. 1b).

Recordings from the near-wave field related to the shock of $M = 2$ ($E = 4 * 10^5$ J) with the hypocentral distance of $R = 210$ m have a similar pulse-shaped character. Here maximum velocities of the ground motion reach the values of the order of 0.01 m/s.

The results of the recordings from the 22E longwall area of "Szombierki" mine are summarized in Table 1.

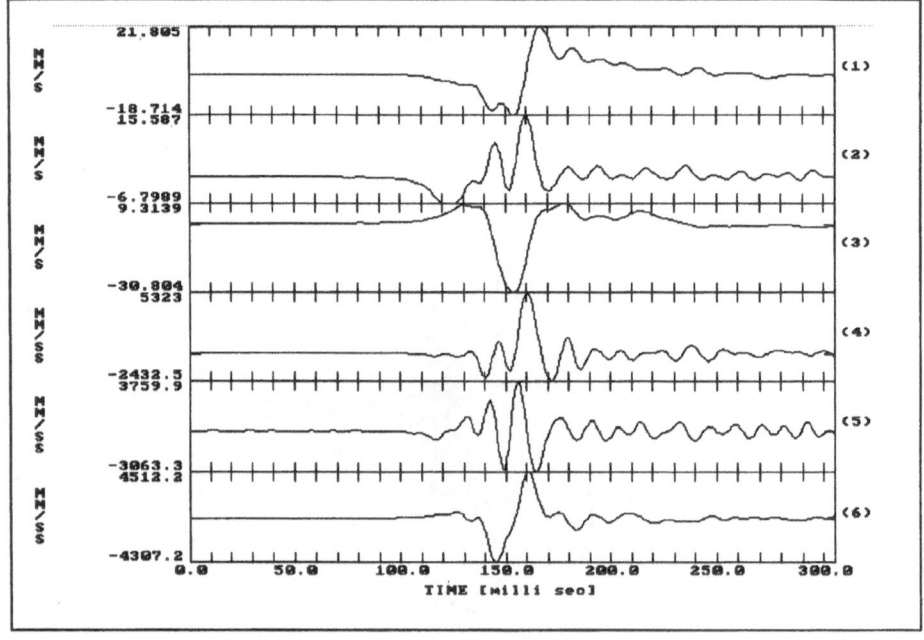

Figure 2a
Velocity (1, 2, 3) and acceleration (4, 5, 6) for a tremor of magnitude $M_L = 2.2$ recorded in a near field (distance $R = 300$ m).

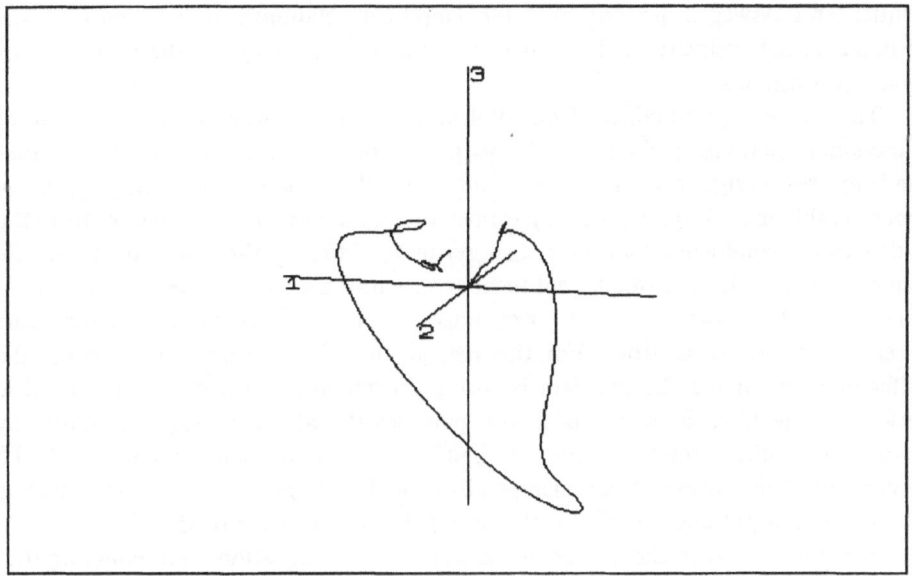

Figure 2b

Particle ground motion (velocity) for a tremor of magnitude $M_L = 2.2$ recorded in a near field (distance $R = 300$ m).

Indirect Methods of Defining the Particle Ground Motion in the Near-wave Field

The number of seismograms with nonclipped recording is, unfortunately, very limited. Butler and Aswegen (BUTLER and ASWEGEN, 1993) provide, for the gold mines in South Africa, some data from which it follows that particle ground velocity in the near field can vastly exceed the value of 10 m/s for strong events.

Table 1

The results of the recording of underground mining tremors in the near-wave field

Magnitude	Energy J	Hypocentral distance m	Z component mm/s	N-S component mm/s	E-W component mm/s
2.2	$1 * 10^6$	300	21.7	15.5	30.8
2.0	$4 * 10^5$	210	6.2	4.2	8.5
1.5	$1 * 10^5$	500	8.3	5.1	8.4
2.0	$4 * 10^5$	250	20.0	6.0	9.0
1.4	$3 * 10^4$	430	1.0	2.2	1.4
1.4	$3 * 10^4$	390	2.0	2.0	2.4
1.5	$4 * 10^4$	450	1.4	2.2	1.9
1.6	$6 * 10^4$	310	1.4	0.4	2.3

Butler and Aswegen present also the empirical relationships to determine the ground particle velocity in the near-wave field, vs. the energy of the event and its epicentral distance.

The recordings obtained from Polish coal mines show that this form of functional relationship does not always prove correct. Underground observations and rare recordings reveal that the tremors with the same seismic energy can differ considerably in rock-mass motion parameters. At the same time, it follows from the calculations conducted that they also seriously differ in the stress drops in the source. Tremors having small sized focal areas with high stress drops are more often the cause of destructions in the workings as compared to those with the same energy and low stress drops. For this reason, the trials are made to alleviate the difficulties in solving the problem by using the relationship between the particle velocity in the focal zone and the source parameters that are feasibly determined by means of a direct interpretation of digital seismograms from the far field. To accomplish this, Brune's relationships (BRUNE, 1970) have been used to calculate the maximum particle velocity in the near field vs. the stress drop.

Accordingly, when the source of the tremor is a dislocation, and assuming that the stress drop on its surface occurs immediately (Brune's model), the stress pulse generates a pure shear wave that propagates perpendicularly to the surface of the focal plane. The initial velocity of particles in the focal area can then be expressed by the realtionship (BRUNE, 1970):

$$V = \frac{\Delta\sigma}{\mu} * C_s \qquad (1)$$

where V—particle velocity; $\Delta\sigma$—stress drop (also called the effective stress, the difference between the initial stress and the kinetic friction level on the fault) determined from the zero-frequencey level, Ω_0 and (generally) the corner frequency, f_c of the far-field displacement amplitude spectrum; μ—modulus of rigidity; C_s—velocity of propagation of shear wave. The time variability of the initial velocity near the dislocation surface may, in turn, be described by Brune's formula (BRUNE, 1970):

$$V(x = 0, t) = \frac{\Delta\sigma}{\mu} * C_s * \exp\left(\frac{-t}{\tau}\right). \qquad (2)$$

In this approximation: $\tau = (r/C_s)$, r—the approximate radius of an equivalent circular dislocation, t—time.

Calculations of particle ground velocity are performed for the seismic events recorded at distances extending 500 m from the source.

From the conducted investigation it follows that in the case of the near-wave field, the seismic energy is an important but not fundamental factor from the point of view of the values of dynamic vibration parameters. The stress drop $\Delta\sigma$ proves to be the above fundamental factor.

Table 2

Particle ground motion and dynamic stresses from mining tremors near the focal zone

Energy, J	$\Delta\sigma$, MPa	V, m/s	σ_{xy}, MPa	σ_x, MPa
$2 * 10^6$	1.8	0.3	1.8	2.9
$3 * 10^8$	3.64	0.6	3.2	5.8
$1 * 10^6$	1.18	0.22	1.2	2.1
$2 * 10^9$	10.0	1.6	9.6	15.4
$3 * 10^3$	30.0	5.5	30	51.4

In practice the situation is often encountered during which a tremor with rather low energy, of the order of 10^5 J, but with considerable stress drop and the relatively small size of the focal area is considerably more dangerous, from the point of view of its impact on mine workings, compared to an event with higher energy but with less stress drop.

Evaluation of seismic Loads Influencing the Mine Workings

The wave field transporting the vibrational motion energy of the medium, generated by a mining tremor causes additional dynamic (seismic) loads within the structural elements of the medium which are added to the existing static stress. The possible evaluation of these loads exist if the values of particle velocity, velocity of propagation of seismic waves and geomechanical properties of the medium are known.

In the event that a seismic wave propagates in a homogeneous and isotropic medium along x direction, the resulting dynamic stresses are given by the relationship (BRADY and BROWN, 1985):

$$\sigma_x = \rho C_P V_x \tag{3}$$

$$\sigma_{xy} = \rho C_s V_y \tag{4}$$

$$\sigma_{xz} = \rho C_s V_z \tag{5}$$

$$\sigma_y = \sigma_z = \sigma_x \frac{v}{1-v} \tag{6}$$

where: σ_x, σ_y, σ_z —normal stresses; σ_{xy}, σ_{xz} —shear stresses; ρ —density; v —Poisson's ratio.

The values of calculated parameters, namely particle ground velocities and sudden increases of dynamic stresses near the focal zone are given below. The values of stress drops were calculated for selected mining tremors from the Upper Silesian Coal Basin (Table 2).

There are empirical scales giving relationships between the values of particle velocity and the effect of their impact on the structure of the rock medium (i.e., Dowding and Rosen scale, in JOHN and ZAHRAH, 1985). Lack of damage was observed in the case when particle velocity was below 0.2 m/s, and well-marked damage (large spalling) occurred when particle velocity was above 0.4 m/s.

This means, in relation to mining tremors, that the events characterized by stress drops of the order of 1.5–2.0 MPa can be dangerous for the structure of rib strata of mine workings and should be taken into account from the point of view of the stability of underground mine openings.

Examples of Application of Near-source Parameters for Formulation of Rockburst Conditions in a Mine Gateroad

Recently, the largest rockburst problems in the Polish mining industry were generated by roof and roof-bed rockbursts (i.e., bed rockburst initiated by a seismic shock from the roof). Statistics from the last several years illustrated that the roof-bed rockbursts occurring in gateroads are the most dangerous and their effects are, at the same time, very difficult to predict.

Apart from the basic factors causing rockbursts of this kind, i.e., very high pressure at great depths and the natural liability of coal to dynamic disintegration under loading, the defined field of dynamic stresses, initiated by roof tremors, is a very important factor (initiator for the coal burst) in this process. The possibility of determining the particle motion and dynamic stress field close to the source has been described in the previous chapters.

Ascertaining the effects of all those factors, suitable models of the rockbursts could be developed.

The rockburst in N°5 longwall at Bobrek mine at 830 m depth has been adopted for analysis. The gateroad was driven in the 6.1 m coal seam in the bottom slice. The rockburst was initiated by a tremor with a seismic moment of $8 * 10^{12}$ Nm ($M = 2.1$), the hypocenter of which was 90 m distant from the working. There was an edge from earlier mining in the rockburst area, and the pressure increase coefficient, determined from the seismic measurements (DUBIŃSKI, 1989) was about 150%, which yields the stress concentration coefficient in the rockburst area, $k = 2.5$. The rockburst caused destruction of a 50 m-long section of the gateroad as well as miners' injuries.

The sudden increase of dynamic stresses has been calculated according to the formula:

$$P_d = W \rho_c C_c V \tag{7}$$

where $W = [2/(n + 1)]$ —ground motion amplification at the rock-coal boundary; ($n = \rho_c C_c / \rho_r C_r$: $\rho_c C_c$ —density and velocity in the coal seam, $\rho_r C_r$ —density and

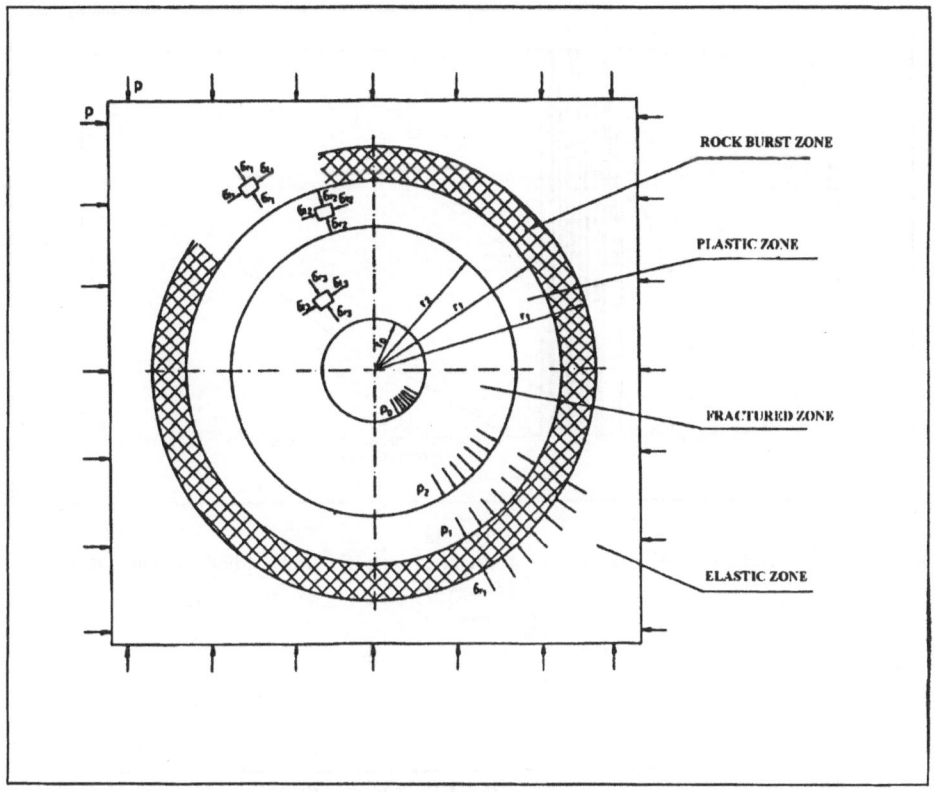

Figure 3
Schematic diagram of the rock-mass model.

velocity in the rock). The particle ground motion V_s has been calculated according to the formula (2), $\Delta\sigma$ has been determined from the scaling relation of strong tremors for the Upper Silesian Coal Basin (MUTKE *et al.*, 1993):

$$\log M_0 = 1.13(\log \Delta\sigma) + 11.1 \tag{8}$$

where $\Delta\sigma$ [ba]—Brune's stress drop, M_0 [Nm]—seismic moment ($14 \geq \log M_0 \geq 11$).

Calculations were made for the model of a circular working situated in a thick coal seam, for which circumferential and radial stresses were calculated according to CIAŁKOWSKI and MUTKE (1994) as well as radii of individual zones (elastic, plastic and cracked ones, Fig. 3). This rockburst model is based on the assumption that it is generated by the rock mass tremor when the pulse-shaped wave reaches the working, causing a pulsed increase of hydraulic pressure (p) of the value p_d.

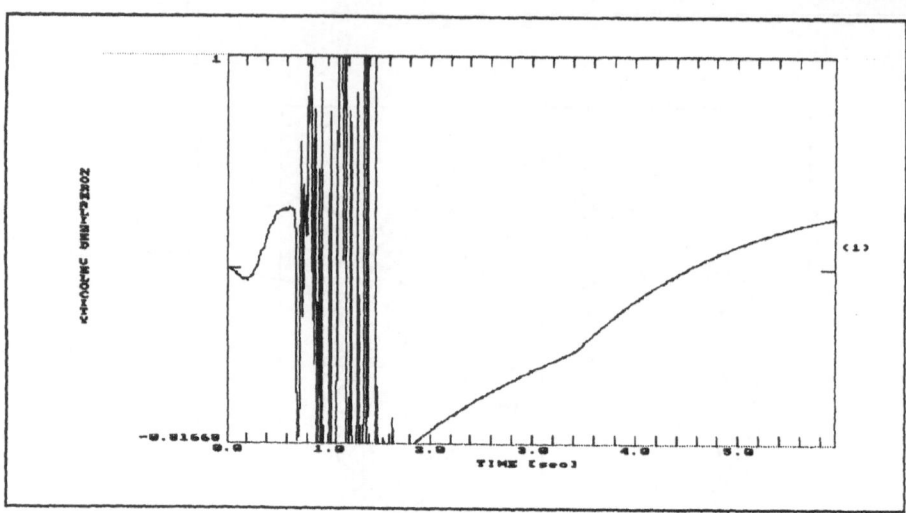

Figure 4a
Records of the rockburst obtained in the destruction zone (Halemba Mine, distance $R = 200$ m, $M_L = 4.1$).

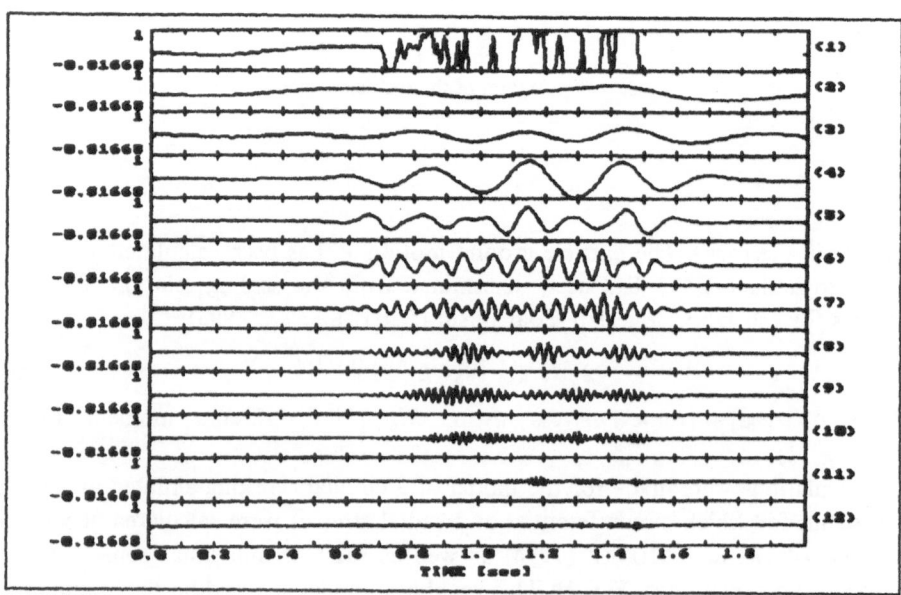

Figure 4b
Velocity seismogram after being band-pass filtered (Halemba Mine, distance $R = 200$ m, $M_L = 4.1$).

After including the dynamic stresses, the balance at the boundary of the elastic and plastic zones is disturbed and can be restored at a radius $r_t > r_1$ (see Fig. 3)

$$r_t = r_1 \sqrt{1 + \frac{p_d}{p - p_1}}. \tag{9}$$

According to MINH (1989) the total coal disintegration energy is equal to the unitary energy of the nondilatational strain:

$$E_t = \int_{r_1}^{r_t} \Phi_{fl} \, dv \tag{10}$$

where Φ_{fl} —unitary energy of the nondilatational strain,

$$E_t = \frac{1.5\pi}{E_1}(p + p_d + p_1)^2 r^4 \left(\frac{1}{r_1^2} - \frac{1}{r_t^2} \right) \tag{11}$$

and the kinetic energy is equal to

$$E_k = E_t \left(1 - \frac{E_1}{M} \right) \tag{12}$$

where E—elastic modulus, M—drop modulus (real slope of post-failure stress-strain curve). The following results were obtained: $p = 48.35$ MPa, $p_d = 1.62$ MPa, $r_1 = 10.9$ m, $r_t = 11.24$ m, $E_k = 5.76$ MJ. This means that the dynamic pulse of the value $p_d = 1.62$ MPa could initiate, under conditions of this working, disintegration of coal within a ring confined by the radii r_1 and r_t and with the kinetic energy of coal disintegration, $E_k = 5.76$ MJ.

Figure 4a shows the record by a seismometer located in the destruction zone. After recording the fragment of oscillations, the seismometer became damaged. Unfortunately, the record has clipped, nonetheless it is possible to evaluate the frequencies of the dominant amplitudes. The seismogram was filtered with a zero-phase band-pass filter in the frequency ranges of 0.5–2; 1.5–3; 2–5; 4–10; 9–20; 19–30; 29–40; 39–50; 49–75 Hz. Analysis of filtering reveals that the range 2–40 Hz was dominant. Time rise velocity is comparable on this record with the time of coal disintegration (KIDYBIŃSKI, 1982). This can testify that, under the Polish conditions, this frequency range is important from the point of view of coal disintegration (an explosive rockburst), taking place after the arrival of the pulse from the roof tremor. Such frequencies are typical for the distance of 50–500 m from the seismic source. It seems that the pulse with very high frequency (more than 100 Hz), compared with the coal disintegration time (being under the Upper Silesian conditions in the range 0.1–0.01 s), should not cause an explosive disintegration of coal. This disintegration is likely to be induced by vibrations with large amplitudes and low frequencies, resulting then as a source of high loads. The rockburst at Miechowice mine, in the gateroad, serves as an example. The rock-

Table 3

The permissible acceleration values depending on the pulse duration time for standing persons

Direction of acceleration vector	Position	Notes	Pulse rise velocity m/s	Duration time s	Acceleration m/s^2
Horizontal	Standing	Holding on to the balustrade	1.8	0.6	3.0
Horizontal	Standing	Free	0.8	0.7	1.1
Horizontal	Standing	Free	3	0.12	25.0

mass tremor was localized at a short distance from the gateroad. Miners' injuries (including fatalities) were sustained as the result of rock-mass motion. There was no injury caused by coal entombment and suffocation of men, and the gateroad suffered only a partial destruction. Consequently there was the possibility of free passage. In the Polish mining industry such events are frequent.

A detailed analysis of the official medical records (DUBIŃSKI and LIPOWCZAN, 1994) enabled us to determine the main injuries in the organisms of the six dead miners:

- skull-face injury (6 casualties)
- lungs injury (4 casualties)
- heart injury (3 casualties)
- liver injury (4 casualties)
- total breaking of the spine (5 casualties)

The range of the resonant frequencies for the human body and organs is 3–25 Hz (e.g., head 4 Hz and 17–25 Hz, liver 3–4 Hz, spine 8 Hz, man sitting 5–12 Hz, man standing 4–6 Hz). These are just the frequencies typical for the seismic near-source wave field (see Fig. 4b). The permissible acceleration values depending on the pulse duration time for standing persons are presented in Table 3 (GLAISTER, 1978).

Conclusions

1. The conducted study and observations of excavations stability (particularly of coal headings) as well as the analysis of miners' injuries demonstrate that to gain improvement in this field the knowledge of real vibration parameters from the near-source wave field is indispensable.

2. A trial to compensate a serious lack of *in situ* recordings from the near field can be made either through establishing the local-range empirical relationships or through relationships of scaling of the source parameters for a specific area.

3. Seismograms of the near-wave field show the pulse shape which can be related to a single exciting force of a shock character.

4. In recent years in Polish coal mining, the rockbursts in mine galleries proved to be the most severe in their consequences. These were the roof-bed rockbursts, triggered off in the bed by a stress pulse from the nearby roof tremor. The computational example proves that in a thick coal seam, a pulse with the stress $\sigma = 1.8$ MPa from the tremor with $M = 2$ and at a distance of $R = 90$ m, generates a rockburst zone limited by the radii $r_1 = 10.9$ m and $r_t = 11.24$ m, with the kinetic energy of the emitted coal of $E_k = 5.76$ MJ.

5. The results of the mining-shock analyses point out that the following vibration parameters can be expected close to the dislocation zone:

- velocity amplitude: 0.01 m/s–10 m/s
- sudden increases of the stress field: 0.1 MPa–50 MPa

These parameters are sufficient to cause the death of miners present in the shock zone (expelling the bodies into the surrounding space).

REFERENCES

BUTLER, A. G., and ASWEGEN, G. (1993), *Ground Velocity Relationships Based on Large Sample of Underground Measurements in Two South African Mining Regions*, 3rd International Symp. on Rockburst and Seismicity in Mines, Kingston, Ontario, Canada, Proc. A.A. Balkema, Rotterdam, pp. 41–49.

BRADY, B. G., and BROWN, E. T., *Rock Mechanics for Underground Mining* (George Allen and Unwin, 1985) 527 pp.

BRUNE, J. N. (1970), *Tectonic Stress and the Spectra of Seismic Shear Waves from Earth-quakes*, J. Geophys. Res. *75*, 4997–5009.

CIAŁKOWSKI, B., and MUKTE, G. (1994), *An Influence of the Support Bearing Capacity on the Extent and Effects of the Rockburst Occurrence in a Passageway Opening* (in Polish), Sc. Works of the Geotechnics and Hydroengineering Inst. of the Wrocław Technical University, Wrocław, No. *65*, 23–32.

DUBIŃSKI, J. (1989), *Premonitory Assessment of Mining Tremors Hazard in Coal Mines by Seismic Method* (in Polish), Sc. Works of the Central Mining Institute, Katowice, 163 pp.

DUBIŃSKI, J., and LIPOWCZAN, A., *The analysis of the effect of rockbursts influence on miners* (in Polish). In Proc. of Conf. *Rockbursts '94 – Engineering Solutions of Rockbursts Problems* (Central Mining Institute, 1994) pp. 135–147.

GIBOWICZ, S. J. (1989), *The Mechanism of Seismic Events Induced by Mining* (in Polish), Pub. of the Inst. of Geophysics Pol. Acad. of Sc., M–13 (221), PWN-Warszawa, 106 pp.

GLAISTER, D. H. (1978), *Human Tolerance to Impact Acceleration Injury*, The British J. of Accident Surgery *9*.

JOHN, C. M., and ZAHRAH, T. F. (1985), *Aseismic Design of Underground Structures*, prepared under National Science Foundation, Agbabian Associates, El Segundo, California.

KIDYBIŃSKI, A. (1982), *Principles of the Mining Geotechnics* (in Polish), Ślask-Katowice.

KOSTROV, B. V., and SHAMITA DAS, *Principles of Earthquake Source Mechanics* (Cambridge University Press, 1988).

MINH, V. C. (1989), *Energy Analysis of Deformation and Failure of Rocks*, Ph.D Thesis, University of Warszawa, No. 436/89.

MUTKE, G. *et al.*, (1993), *Methods of Determining Source Parameters of Mining Tremors in Mines* (in Polish), Report of Investigations, Central Mining Institute, Katowice, Poland.

(Received December 28, 1994, revised October 12, 1995, accepted October 16, 1995)

PAGEOPH, Vol. 147, No. 2 (1996)

0033-4553/96/020263-14$1.50 + 0.20/0

Induced Seismicity in Large-scale Mining in the Kola Peninsula and Monitoring to Reveal Informative Precursors

N. N. MELNIKOV,[1] A. A. KOZYREV,[1] and V. I. PANIN[1]

Abstract — Large volumes of rock mass, mined-out and moved within these deposits, resulted in irreversible changes in the geodynamic regime in the upper earth's crust of the adjacent territory. These changes manifest themselves in a more frequent occurrence of such intensive dynamic phenomena as tectonic rock bursts due to fault movement adjacent to the area which is mined-out and man-made earthquakes which sharply decrease mining safety and result in great material losses.

To develop the prediction techniques of such phenomena, a monitoring system is created, based on the program of the Kola Complex of geodynamic measuring stations. Most of this system is realized in the region of the Khibiny apatite mines. The system provides regional seismological monitoring, local prediction of seismicity in separate areas of a rock mass and, determination of stress and strain in rock masses, local geophysical monitoring over the state of rocks in a rock mass as well as physical and mathematical modelling of geodynamic processes in the upper earth's crust.

The investigations have resulted in the distinguishing of some regularities in manifestations of induced seismicity and tectonic rock bursts and in the determination of strain precursors of intensive seismic events in the Khibiny mines.

The mechanism is provided by the induced seismicity which resulted from the anthropogenic impact on the geological medium. A geodynamic monitoring complex is described, which is used to reveal the precursors of powerful seismic events *in situ*, and monitoring results are shown, obtained in the Kola Complex of geodynamic stations. Methods of preventing tectonic rock bursts and induced earthquakes are presented.

Key words: Seismicity, rock bursts, stresses, mining operations, monitoring, geodynamic regime, informative precursors.

Introduction

The notion "induced earthquakes" is widely used in seismology for seismic events due to filling the reservoirs with water or from exploitation of enormous oil and gas deposits (NIKOLAYEV, 1988; GUPTA and RASTOGI, 1979). Such excited, induced earthquakes sometimes are rather powerful and attain the magnitude M6, which results in tragic consequences. Such events (increase in seismicity of up to 10

---

[1] Mining Institute, Kola Science Centre, Russian Academy of Sciences, Apatity, Murmansk Region, 184200, Russia.

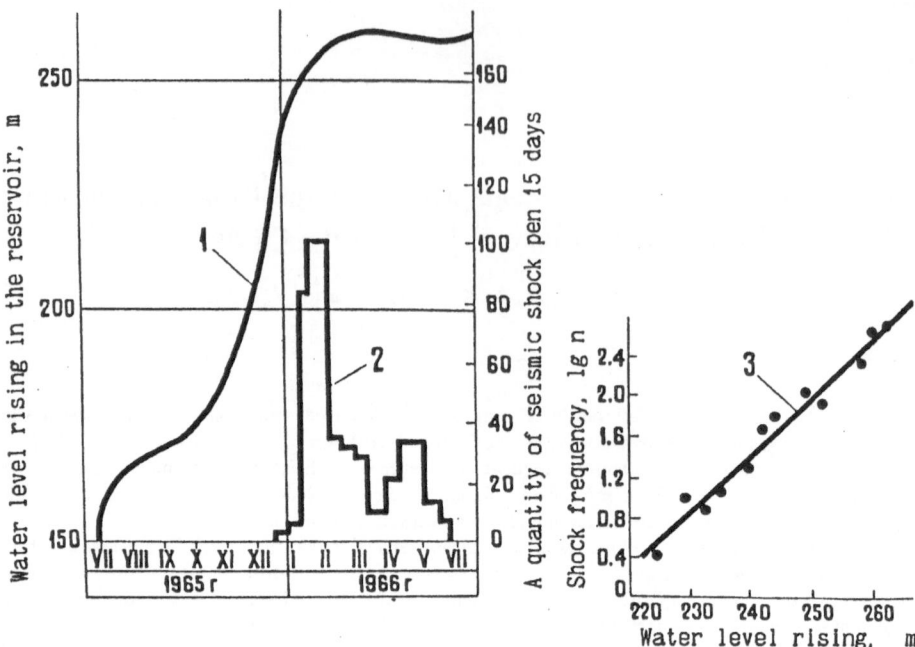

Figure 1
Change in seismicity when the reservoirs are being filled with water. Increase in the water level (1) and
in seismic activity (2) and the dependence of a quantity of shocks on the water level (3) in the Kremasta
reservoir (GUPTA and RASTOGY, 1979).

times) were observed in the territory of the former USSR when filling the reservoirs
of the Nurek, Toktogul, Charvak, Inguri, Chirkeiskaya, Zeiskaya and other hydroelectric power stations (NIKOLAYEV, 1977; BABAYEV *et al.*, 1977; ULOMOV and
BEZRODNY, 1977; LEVKOVICH and IDAIMACHYEV, 1977; KUZIN and KALININ,
1981). In some hydroelectric power stations an increase in seismicity was observed
to be directly independent on the rise of water level in the reservoir (Fig. 1).
Induced seismicity resulting from the exploitation of oil and gas deposits was
registered in the U.S.A. (the Rangley and Denver deposits) (GUPTA and RASTOGI,
1979), in Uzbekistan (the Gazli earthquakes, 1976 and 1948) (PLOTNIKOVA *et al.*,
1990), in Dagestan (the Starogroznenskoye earthquake, 1971, Oktyabr'skoye, 1973)
(SMIRNOVA, 1977).

An increase in seismicity is normally accompanied by a change in the region's
geodynamic regime. For instance, when the Kariba reservoir in Rhodesia was filled
with water, the total bending of its bottom reached 12 cm, which was conrfimed by
high-precision levelling. When the Groznenskoye earthquakes occurred, the maximum lowering of the surface was equal to 13–15 cm. It is significant that the zones
of highly induced seismicity are usually characterized by intensive recent tectonic

movements and by compressive horizontal stresses of substantial values (40–60 MPa), recorded in the upper crust.

Accordingly, seismic events are also recorded in large-scale mining when both ore and coal deposits are mined out (AITMATOV, 1987; JOHNSTON, 1989). Tectonic rock bursts and induced earthquakes resulting from mining are the questions under discussion. These events have resulted from high regional stresses and movements along different tectonic fractures which appeared as a result of mining.

Most powerful events similar to induced earthquakes have been recently recorded in Germany in potassium deposit Werra, Thuringia ($M \sim 5.5$, 13.03, 1089) (KNOLL, 1990), in the Ostravo-Karvinsk coal basin, Slovakia ($M \sim 3.1$–4.5, 27.04, 1983) (KONECNY, 1989), in the Kirovsk mine, the Apatit Company, Khibiny, Russia ($M \sim 4.2$–4.3, 16.04, 1989; $M \sim 3.0$, 25.07, 1989) (SYRNIKOV AND TRYAPITSYN, 1990), in the northern Ural and southern Ural boxite mines ($M \sim 3.0$, 5.10, 1984; $M \sim 3.5$–4.0, 28.05, 1990) (KOLESOV, 1993; LOMAKIN and JUNOSOV, 1993), in the Tashtagol iron-ore deposit, Gornaya Shoria ($M \sim 3.0$–3.5, 12.08, 1984) (LOMAKIN and JUNOSOV, 1993). A list of deposits where rock-tectonic bursts and induced earthquakes are recorded is presented by AITMATOV (1987) and JOHNSTON (1989). According to the provided data, tectonic rock bursts are observed in all deposits with anomalous, high, horizontal stress despite the type of tectonic structure (stable shields, moveable platforms, mobile mountain-folded areas). Induced earthquakes are recorded, as a rule, in the deposits where large volumes of rock are mined-out and displaced. In our data (Fig. 2), only a certain volume of mined-out rock disturbs the geodynamic regime in the region, which results in the analogous condition of the induced seismicity similar to that recorded in filling a reservoir with water.

A Model of the Process

In our opinion, tectonic rock bursts may be considered as foreshocks or aftershocks of induced earthquakes. The data given in AITMATOV (1987) and JOHNSTON (1989) allow us to reach the conclusion that rock-tectonic bursts are observed in high-stressed rock masses at shallow depths. Such rock bursts are observed in such rock masses even with far less mining-induced impact.

One of the necessary conditions for seismic events is the structural heterogeneity in a rock mass, formerly formed faults in particular in the vicinity where a sudden redistribution of stresses takes place. As we discovered earlier, the faults substantially change the stress distribution in a block medium, which results in a stress concentration or stress relief in some areas (SAVCHENKO and KOZYREV, 1993). The zones of increased stress concentration are, as a rule, the zones of increased seismicity where the induced earthquakes are expected.

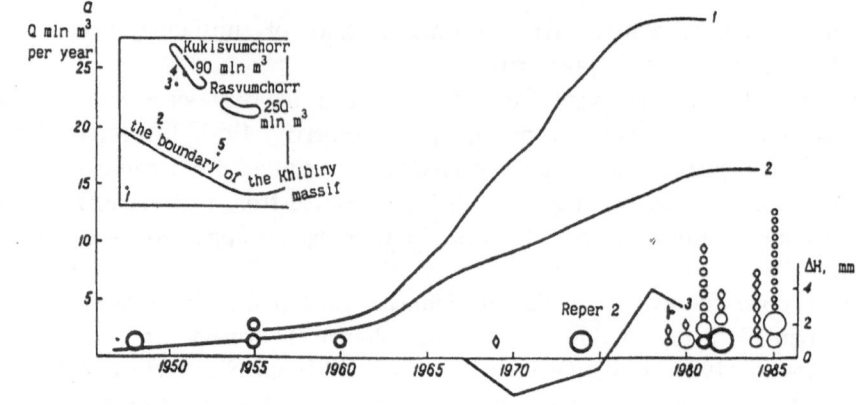

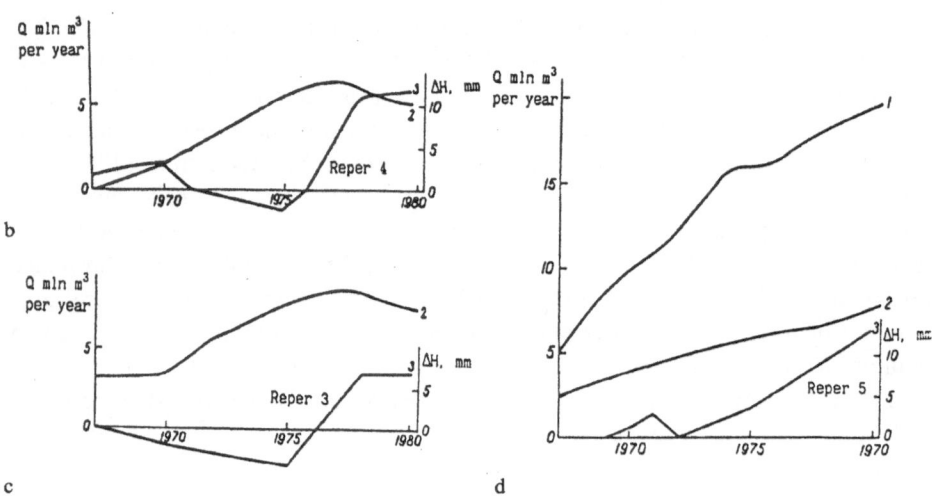

Figure 2

The inducing impact of mining on the geodynamic processes in the Khibiny massif. a—the total rock mass and ore output in the Khibiny mines and geodynamic processes; 1—the total rock mass output through the years; 2—ore output through the years; 3—repers elevation; 1–5—repers numbers; ⊚—earthquakes; ◇—rock bursts; ○—shocks (earth tremor) in a rock mass; ↦—displacement of a rock block; b—ore output in the Jukspor mine and Saami open pit and the repers elevation caused by mining; c—ore output in the Jukspor, Kirovsk mines and in the Saami open pit and the repers elevation caused by mining; d—rock mass output and ore mining in the Central and Rasvumchorr mines and the repers elevation caused by mining.

In accordance with the well-known notions (KOZYREV and PANIN, 1993), induced earthquakes differ from natural earthquakes only in a regime of the seismic energy emitted as a result of man's activity. They are characterized by many foreshocks (or rock-tectonic bursts) occurring before a maximum earthquake, a

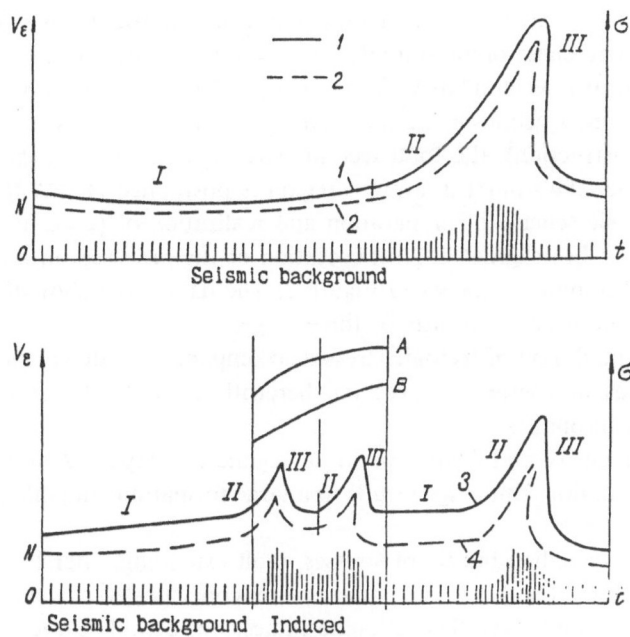

a

b

Figure 3
Scheme of changes in the course of time t, strain V, tectonic stresses σ and seismic energy emission N (KOZYREV and PANIN, 1993). a—under natural evolution of the processes in the long period of time t; b—under the influence of mining; 1—change of stresses σ; 2—change of strain rate V; 3–4—change in stresses and strain rate when certain volume are of mined-out rock masses; I—stage of fracture preparation; II—stage of foreshocks; III—the occurrence of the main shock of the earthquake, redistribution of stresses and manifestation of aftershocks; A—change in the volume of mined-out rock mass; B—change in the stress concentration caused by rock mass mined-out; ——change in time scale and manifestation of induced seismicity.

slow decrease in the quantity of aftershocks, and increased power due to a near-surface seismic center location. Induced earthquakes originate when a number of conditions coincide in mining. They are: large-scale mining (the area mined, its depth, the volume of mined-out and transported rock), the impact of explosion used in mining, the presence of high-strength brittle rocks in a tectonically heterogeneous mining zone, a favorable geomorphological landscape (mountain relief), high horizontal tectonic stresses in a rock mass, suitable tectonophysical conditions (i.e., the zones characterized by high-gradient rates of the latest tectonic movements). These factors result in the crustal stress redistribution, which is in turn reflected in the behavior of separate rock blocks which are slowly creeping or observed in shock waves (shocks) along the faults. This can be accompanied by geodetic, seismic, electromagnetic and other anomalies.

An engineer's activity may be considered as an additional short-term influence on a complicated chain of natural interconnected processes which act as a trigger to release the former accumulated elastic energy. Due to the energy emitted in the vicinity of a mine's geomechanic space (the zone of influence of mining operations or of any construction), the total tectonic stresses decrease and the moment of a possible earthquake's burst in a given region is postponed (KOZYREV and PANIN, 1993). The basic scheme of preparation and realization of powerful seismic events in a rock mass, both under natural evolution of the processes in a rock mass and as a result of mining, is shown in Figure 3. The data given show that the natural evolution of the process consists of three stages:

1) the accumulation of tectonic stresses accompanied by slowly increasing strain. Small portions of energy are observed herewith as weak shocks or earthquakes (aseismic background);

2) stress intensity and change in strain regime accompanied by foreshocks and faults amalgamation; this stage results in a large formation culminating in the main earthquake's shock;

3) subsequent redistribution of stresses, fault extending, aftershocks occurrence.

When mining (Fig. 3b), the stresses are concentrated in the vicinity of a mined-out space and reach their ultimate values considerably earlier and more often in comparison with the natural state. These are observed as weak shocks. Powerful explosions result in additional disturbances in stress fields, which induces the faults to occur much earlier. Therefore, to predict such dangerous phenomena like induced rock bursts and earthquakes, it is important to determine possible deformations and additional stresses induced by large-scale mining, conditions for the liberation of energy accumulated by fracturing along the faults or the fracturing of rocks emplaced between them. Of extreme importance is a study of a seismic regime, of geophysical field variations, crustal deformations using (geodetic) observations, strainmeters and tiltmeters of different systems, monitoring over the change in the state of stress while mining.

The Techniques of Monitoring and Preliminary Results

Monitoring is carried out in accordance with the Programme developed by the Kola Complex of geodynamic measuring stations in apatite and rare-metal deposits mining in the Khibiny and Lovozero rock masses (KOZYREV and IVANOV, 1993).

Monitoring is carried out in five basic directions:

1. stress and strain monitoring;

2. regional seismic monitoring;

3. local monitoring to determine the seismicity in the mines and the horizon's vicinity;

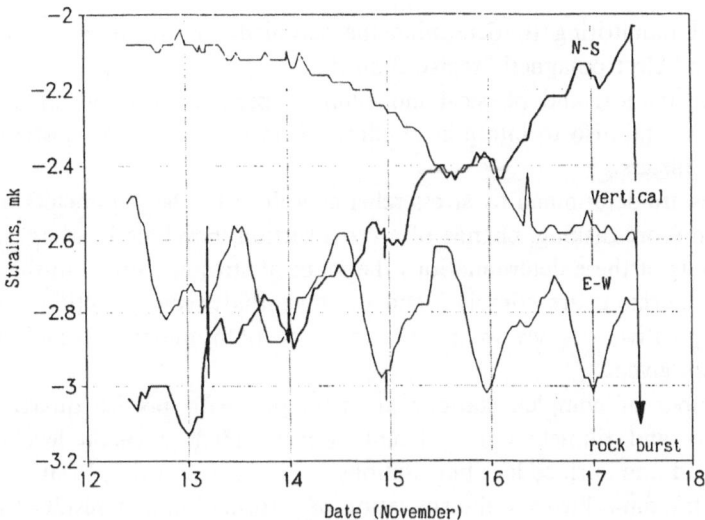

Figure 4
The nature of rock-mass deformation before the earthquake on 17.11.93 by the data of the KD-3
deformometer.

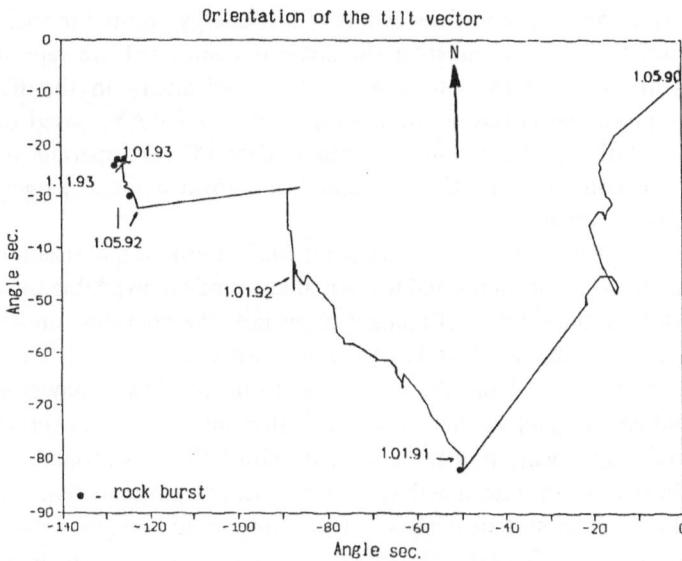

Figure 5
The change in the orientation of the greatest tilt *in situ* in the Kirovsk mine by the NSO tiltmeter data.
The rock burst is shown.

4. local monitoring to determine the seismicity in the mine vicinity by the acoustic and electromagnetic registration data;

5. mathematical and physical modelling to predict the anticipated stress and strain values and also to interpret abnormal fluctuations of the registered physical fields' parameters.

Stresses are determined by stress relief as well as by the parameters of borehole destruction (core disking, change of a cross section, borehole bending). According to the results of these determinations the maps of stresses formed in the vicinity of mines and horizons are compiled and the magnitudes and directions of principal peak compressive stresses in a rock mass of both mined-out and perspective deposits are given.

A number of complex stations are equipped with precise quartz and laser tensometers and tiltmeters of different systems. High-precision levelling at the underground and surface levelling stations is periodically carried out. Tensometric and tiltmetric data illustrate the activation of deformation as a result of man-made factors and the anomalies in the behavior of deformation precursors before powerful seismic events, which is qualitatively shown in Figures 4 and 5.

Regional seismological monitoring has been carried out in Northwest Russia by the Kola Science Centre since 1956*. At present there is a network of stations in Amderma, Apatity, Polyarnye Zori, Polyarny Krug, near the settlement of Pudozha, Arkhangelsk, and in the Spitsbergen as well. First investigations of induced seismicity in the Khibiny have disclosed that present home-made equipment does not allow correct registration of the events of low energy and the precise evaluation of their parameters because most of the stations mentioned are remotely located from the Khibiny. Therefore, to study induced seismicity in the Khibiny, the Norwegian high-precision seismic monitoring system ARRAY, based on a SUN-4 SPARC work station and personnel computers IBM PC AT, operates near the city of Apatity. All computer devices are united and form a local network for data collection and treatment.

The computers are linked with national and international seismological networks and data bases through satellites. An analysis of the available seismic data in Khibiny evidences, that most earthquakes occur near the operating mines and in the southern part of the massif where large tailing storages are located. The tailings are transported to the massif from the plants and from the electric power station, i.e., where the induced impact on the surface is rather intensive (TRYAPITSYN, 1993).

To determine the location of seismic events within the area around the operating mines and horizons, an automated system of seismicity monitoring (AMSSM) *in situ* is developed. The system consists of as many as 48 channels with the upper frequency level up to 250 MHz, the dynamic range to 80 dB, and the energy level

---

* The investigations are carried out by the Kola regional seismological centre of the KSC, RAS.

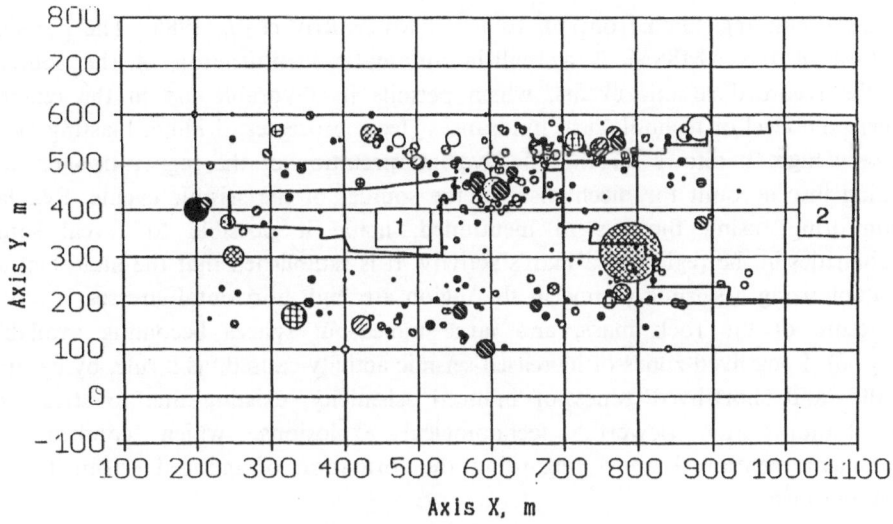

Events of E > 1.0E+4

a

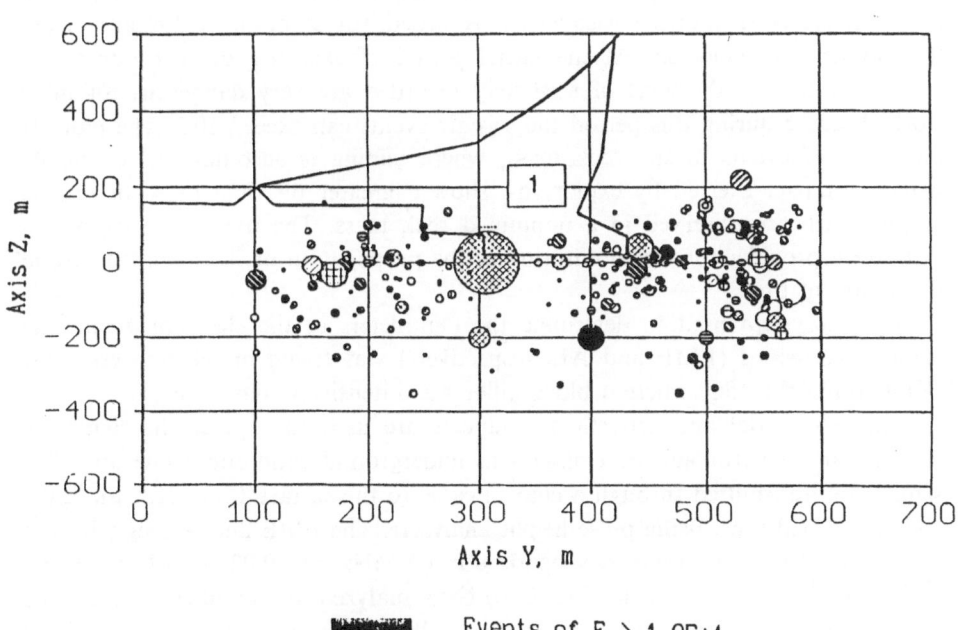

Events of E > 1.0E+4

b

Figure 6
The seismicity by the AMSSM system. a) in a horizontal plane; b) in a vertical section. 1–2—mined-out areas.

of the events registered from 10^2 to 10^7 J (MELNIKOV *et al.*, 1985). The peculiar feature of the AMSSM is a built-in automatic identification of the sources of the recorded seismic events, which permits its favorable use in the regions where the level of technological hindrances (tech. processes: drilling, blasting, etc.) is very high. It allows for the effective interpretation of the registration results, taking into account the mechanism of the sources of the seismic events. Regular monitoring, using the system mentioned, made it possible to reveal some regularities in the region of a man's activity. It is established that the main factors characterizing a seismic regime in the region are high horizontal stresses, a block structure of the rock mass, and large mined-out spaces becoming available (Fig. 6). Long-lived zones of increased seismic activity caused, as a rule, by natural faults, and short-lived zones of induced seismicity, existing due to stress redistribution after powerful technological explosions, which confirm the statement mentioned above regarding the formation of induced seismicity are distinguished.

It is established that induced seismicity increases and decreases in a cyclic manner, with the periods of duration varying and the amplitude reducing in a stable manner. The total duration of induced seismicity substantially depends on the state of stress of rock masses in the zone adjacent to the site of the explosion. When the explosions are carried out in nondisturbed rocks of the hanging wall, i.e., in the most stressed zone, induced seismicity is recorded, sometimes over 2 months from the moment of explosion. At the initial period of existence of such zones, the absolute values of the level of seismicity recorded are very dangerous for mine works because during this period the seismic events can exceed 10 J. The plots of repetition of events in the rock mass, where mining is accompanied by highly dynamic impact exerted by explosions, show a greater tilt than those of seismic events which are observed in a nonmined rock mass. The use of the subsystem mentioned provides for the monitoring of the propagation of the main fracture in the course of time.

The subsystem used to determine the parameters of the electromagnetic and acoustic emissions (EME and AE, respectively) was tested in the Kirovsk mine (Khibiny) in the zone where a block-pillar was intensively mined-out.

Magnetic aerial and acoustic transducers are used to register the signals of acoustic and electromagnetic emission in underground conditions. The amplified signals are transmitted through a coaxial cable to the surface to analyze the EME and AE in real time, using pulse-height analyzers. The EME and AE signals have been registered in the frequency bands $0.1-1.5$ MHz and $0.02-0.1$ MHz, respectively. The following characteristics have been analyzed: the number of EME and AE signals per one measuring cycle; their full and average energy; the mode of power emission; the ratio of full to average energy of the EME and AE signals. The most informative parameters predicting powerful seismic events is a full energy of AE and EME signals and their ratio (Fig. 7).

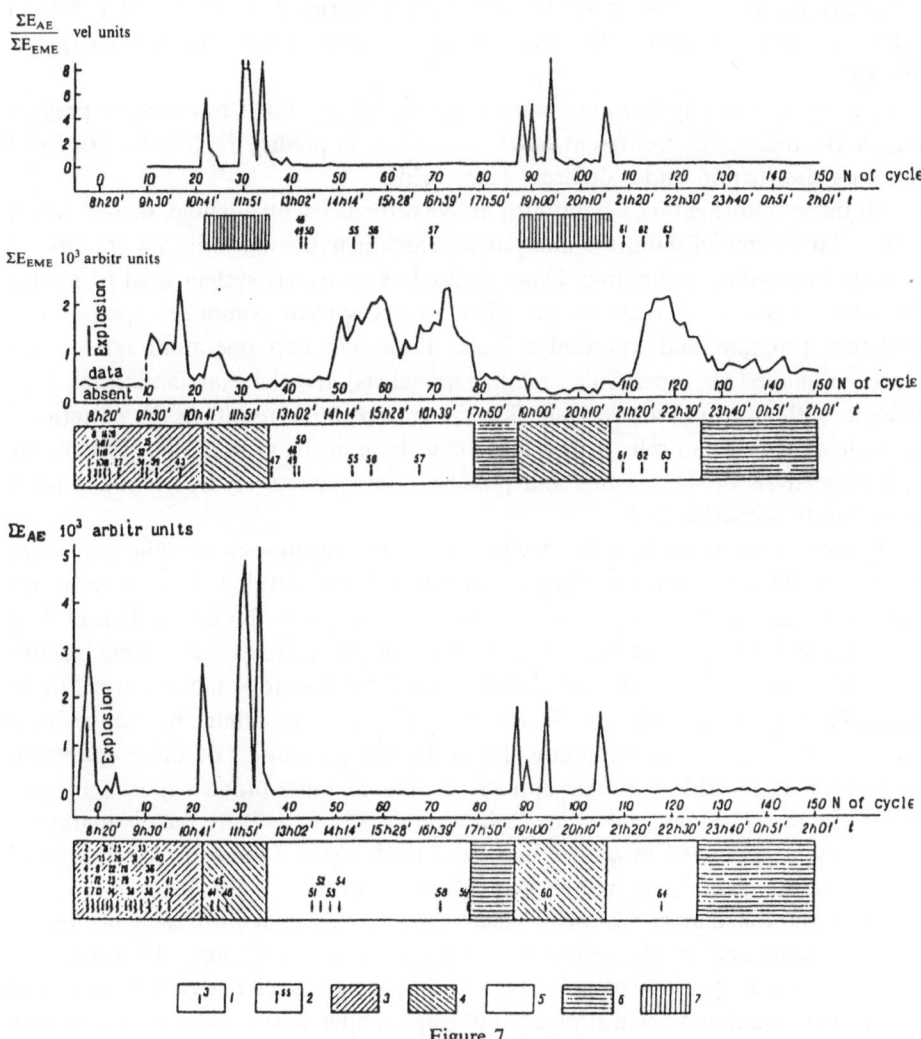

Figure 7

The EME and AE energy values recorded after technological explosion in the Kirovsk mine. 1—seismic events ($E - 10^3$ J); 2—seismic events ($E - 10^4 - 10^5$ J); 3—a stage of rock-mass destruction after technological explosion; 4—a stage of intensive seismological events preparation ($E - 10^4 - 10^5$ J); 5—a stage of occurrence of intensive seismic events ($E - 10^4 - 10^5$ J); 6—a stage of intensity decrease; 7—an interval of abnormal values $\Sigma E_{AE}/\Sigma E_{EME}$.

The monitoring results have been retrospectively correlated with seismic events registered by the automated local seismic monitoring system in this region. The analysis of the obtained data permitted the distinction of different stages in the redistribution process of the EME and AE signals energy *in situ* after a technological explosion due to seismic events of a certain energy class. From these data, a criteria is given to predict the dynamic events characterized by the energy of 10^4-10^5 J.

A subsystem of physical and mathematic modelling allows to calculate preliminary stress and strain at different stages of mining, to predict the behavior of rocks and their destruction under different force fields.

All the monitoring types mentioned above provide us, on a whole, with a rather substantial volume of data; the analysis of which is impossible without creation of modern information technology being realized as an expert system used to predict the state of stress of rock *in situ*. The expert system comprises special data treatment program and knowledge base. There are two operation modes: the education mode (for experts, i.e., mining specialists) and the examination mode (a dialogue with an expert-interpreter). The system operation results in the patterns of the state of rock in all cells of the space data base in the examination conclusion, with an analysis of the reasons and possible consequences of the event evolution according to scenarios 1–2.

At present the measures necessary to prevent significant rock-tectonic bursts and induced earthquakes when mining, are not sufficiently developed. There are only suggestions aimed at lowering the seismicity in active faults by fan drilling of deep boreholes to fire a camuflet explosion (creation of a fractured zone by hole blasting to transfer the zone of stress concentration far from the stope face contour) or by hydraulic fracturing with the formation of a corresponding protective zone (PETUKHOV *et al.*, 1984; Petukhov, 1993). Partial prevention of these dangerous phenomena is possible through optimal mining in burst-prone deposits by pinpointed large explosions or powerful vibration devices to activate the dissipation of the energy accumulated in a rock mass and to decrease a probable occurrence of induced earthquakes in the mines' geomechanic space.

To lower the danger of rock-burst occurrence in structural elements of the mining system and in the entire set of underground workings, the traditional stress-relief methods are used, inferring the creation of special advanced zones and slots in horizontal and vertical planes utilizing an advanced camouflet explosion in boreholes drilled in the most stressed zones of workings, or using specially-lined boreholes between the distance which is 2.0–4.0 of a borehole diameter, depending on the intensity of the stresses *in situ* and on the properties of rocks. The methods mentioned are widely used as preventive measures in the zones of workings characterized by the 1st and 2nd degree of burst danger and in block-pillars mining as well in the Kola Peninsula (JUROV *et al.*, 1989; IVANOV *et al.*, 1989).

Conclusion

In conclusion one more aspect concerning the solution of this problem is purposeful to note. Great attention is paid in a number of home and foreign organizations to developing the methods of tectonic rock bursts prediction. But the actual condition shows that this problem is far from being solved. It is evident that closer international cooperation is necessary to successfully solve this problem. The most real and effective form of such cooperation could be joint studies in the Kola Complex of geodynamic measuring stations where it would be possible to test the latest instruments and equipment, to determine the informative precursors of seismic events of different energy classes, to develop and test the appropriate prediction techniques and preventive measures aimed at lowering the probability of rock-tectonic bursts and induced earthquake occurrences.

Acknowledgment

Studies are carried out with support given by the Russian Foundation for Fundamental Research, Project 93-05-08090.

REFERENCES

AITMATOV, I. T., *Geomechanics of Ore Deposits in Middle Asia* (Frunze, Ilim, 1987) 246 pp.

BABAYEV, A. M., KULAGIN, V. K., LYSKOV, A. M., MAMADALIYEV, JN., MIRZOYEV, K. M., SOBOLYEVA, O. V., STARKOV, V. I., SHKLYAR, G. P., and KHAKIMOV, Induced seismicity in the vicinity of the Nurek reservoir. In *Effect of Engineering Activity on the Seismic Regime* (M., Nauka, 1977) pp. 69–94.

GUPTA, H., and RASTOGY, B., *Dams and Earthquakes* (N. Mir, 1979) 234 pp.

IVANOV, V. I., KOZYREV, A. A., and BELOV, N. I. *et al. Interim instructions on prevention of rock bursts in workings by drilling stress relief boreholes in the mines of the Apatit-Joint-Stock Company. Apatity-Kirovsk* (Publ. House of the Kola Branch of the USSR Academy of Sciences, 1989) 20 pp.

JOHNSTON, I. C. (1989), *Rock Bursts from a Global Perspective*, Gerlands Beiträge zur Geophysik *98* (6) Leipzig, 474–490.

JUROV, A. S., IVANOV, V. I., KOZYREV, A. A. *et al. Development of ways to reduce the danger of rock bursts in rocks when workings are driven using an advanced explosion. In Scientific and Technical-Progress at the Apatit-Joint-Stock Company.* Part I (M., Gighs, 1989) pp. 118–125.

KNOLL, P. (1990), *The Fluid-induced Tectonic Rock Burst of March 13, 1989 in the "Werra" Potash Mining District of the GDR (First Results)*, Gerlands Beiträge zur Geophysik, *99* (3), Leipzig, 239–245.

KOLESOV, V. A., *The state of works on the problem of rock bursts in the mines of the Sevuralboksitruda.* Report presented at the VIII-th coordination meeting on solution of the problem of rock bursts in ore and non-ore deposits mining. Kirovsk, June 25–26, 1991. In *Rock Bursts in Ore and Non-ore Deposits Mining.* (Publ. House of the Kola Science Centre, RAS, Apatity, 1993) pp. 50–57.

KONECNY, P. (1989), *Mining-induced Seismicity (Rock Bursts) in the Ostrava-Karvina Coal Basin, Czechoslovakia*, Gerlands Beiträge zur Geophysik *98* (6), Leipzig, 525–547.

KOZYREV, A. A., and IVANOV, V. I., *Studies on the problem of rock bursts in ore deposits in the Kola Peninsula.* In *Prediction and Prevention of Rock Bursts in Ore Deposits* (Publ. House of the Kola Science Centre, RAS, Apatity, 1993) pp. 18–32.

KOZYREV, A. A., and PANIN, V. I., *Effect of large-scale mining operations on the geodynamic behaviour of the area, and mine-induced seismicity manifestation.* In *Safety and Environmental Issues in Rock Engineering* (A.A. Balkema, Rotterdam/Brookfield, 1993) pp. 841–843.

KUZIN, I. P., and KALININ, N. I., *The peculiarities of the seismicity in the zone of the Inguri Hydro-Power Station's reservoir.* In *Geological-geophysical Studies on the Region of the Inguri Hydro-Power Station.* Tbilisi (Metsniiraba, 1981) pp. 285–296.

LEVKOVICH, R. A., and IDARMACHYEV, Sh. G., *The seismicity in the region of the Chirkeysk reservoir during its filling with water.* In *Effect of the Engineering Activity on the Seismic Regime* (M., Nauka, 1977) pp. 35–37.

LOMAKIN, V. S., and JUNUSOV, F. F., *Effective seismological monitoring in mines.* In *Rock Bursts in Ore and Non-ore Deposits* (Apatity, Publ. House of the Kola Science Centre, RAS, 1993) pp. 73–76.

MELNIKOV, N. N., RASPOPOV, O. M., and JERUKHIMOV, A. Kh. *et al.* (1985), *A New Instrument to be Used in Mining Geophysics,* Vestnik A. N. of the USSR *15*, 6–15.

NIKOLYEV, N. I., *On the state of study of the problem of induced earthquakes caused by engineering activity.* In *Effect of Engineering Activity on a Seismic Regime* (M., Nauka, 1977) pp. 8–21.

NIKOLAYEV, N. I., *The Modern Tectonics and Geodynamics of Lithosphere* (M. Nedra, 1988) 491 pp.

PETUKHOV, I. M., *Scientific and mining-experimental works to solve the problem of rock bursts.* In *Prediction and Prevention of Rock Bursts in Ore Deposits* (Publ. House of the Kola Science Centre, RAS, Apatity, 1993) pp. 11–17.

PETUKHOV, I. M., JEGOROV, P. V., and VINOKUR, B. Sh., *Prevention of Rock Bursts in Mines* (M., Nedra, 1984) 230 pp.

PLOTNIKOVA, L. M., FLYONOVA, M. G., and MACHMUDOVA, V. I. (1990), *Induced Seismicity in the Gazli Gas Field Region,* Gerlands Beiträge zur Geophysik. *99* (5), Leipzig, 389–399.

SAVCHENKO, S. N., and KOZYREV, A. A., *Investigations on stress state near the neighbouring fracture by boundary elements method.* In *Assessment and Prevention of Failure Phenomena in Rock Engineering* (Balkema, Rotterdam/Brookfield, 1993) pp. 251–256.

SMIRNOVA, M. N., *Induced earthquakes as a result of oil deposits development (the Starogroznensk earthquake).* In *Effect of Engineering Activity on the Seismic Regime* (M., Nauka, 1977) pp. 128–141.

SYRNIKOV, N. M., and TRYAPITSYN, V. M. (1990), *On Mechanics of an Induced Earthquake in the Khibiny,* DAN USSR *314* (4), 830–833.

TRYAPITSYN, V. M., KREMENETSKAYA, E. I., and CHEREVKO, V. S. *et al., Effect of large-scale mining on induced seismicity in block rock masses.* In *Prediction and Prevention of Rock Bursts in Ore Deposits* (Publ. House of the Kola Science Centre, RAS, Apatity, 1993) pp. 76–83.

ULOMOV, V. I., and BEZRODNY, Je. M., *Complex instrumental studies of geodynamics near the Charvak reservoir.* In *Effect of Engineering Activity on the Seismic Regime* (M., Nauka, 1977) pp. 29–34.

(Received March 15, 1995, revised October 25, 1995, accepted December 15, 1995)

PAGEOPH, Vol. 147, No. 2 (1996)

0033–4553/96/020277–12$1.50 + 0.20/0

Space-time Interaction Amongst Clusters of Mining Induced Seismicity

A. KIJKO[1] and C. W. FUNK[2]

Abstract—Elementary cluster analysis of induced seismicity in a South African gold mine has shown that there is a clear interaction amongst the clusters; and that the level of the interaction is a function of the distance. The clustering algorithm used is an adaptation of the single-link cluster analysis which considers both three-dimensional space and time. A high level of interaction between the clusters is demonstrated from the cross-correlation analysis of seismic activity rates and radiated energy. A distinct decrease in the value of correlation coefficients was detectable as distance increased. This was somewhat surprising, considering the simplicity of the technique used. Since no attempt is made to study the physical mechanisms of interaction, these results are very preliminary, but interesting from an observational point of view.

Key words: Induced seismicity, clustering, space-time interaction.

Introduction

Extensive studies have been performed on the space-time-magnitude/energy/seismic moment distribution of seismic events, for different regions, and on different scales. Even after the elimination of fore- and aftershocks, the tendency to form nests, swarms, and clusters is often observed. A variety of techniques has been applied in order to quantify the spatial and temporal properties of various catalogs. Some recent investigations include KAGAN and KNOPOFF (1980), KAGAN (1981), DZIEWONSKI and PROZOROV (1984), JOHNSON et al. (1984), NATALE and ZOLLO (1986), ENEVA and PAVLIS (1988). The significant place in the determination of space-time clustering of seismicity belongs to fractal formalism (MANDELBROT, 1982, 1989). Many studies have indicated that the occurrence of seismic events is fractal in space and time and that fractal formalism can be successfully applied to describe the space-time clustering in a wide range of energies, i.e., from acoustic emissions to devastating earthquakes (e.g., SMALLEY et al., 1987; KAGAN and JACKSON, 1991a; LEI et al., 1993; XIE and PARISEAU, 1993). In investigations of space-time anomalies of sesimic activity patterns, significant attention is given to the

[1] Council for Geoscience, Geological Survey of South Africa, Private Bag X112, Pretoria, 0001, RSA.

[2] ISS International Ltd., P.O. Box 2083, Welkom 9460, Republic of South Africa.

detection of seismic quiescence. Several investigations have found that, in some time prior to significant seismic events, a characteristic decrease in seismic activity is observed near the hypocentral region. Comprehensive reviews of such a phenomenon include HABERMANN (1988), WYSS and HABERMANN (1988), REASENBERG and MATTHEWS (1988), WARDLAW et al. (1990) and recently KAGAN and JACKSON (1991b).

Similar studies, relating to the detection of anomalous seismicity patterns and clustering in space-time energy, have only recently been applied to mining-induced seismicity. Investigations have proven that, in many respects, mine-induced seismicity and tectonic seismicity are quite similar and the processes involved in the creation of anomalous seismicity patterns should be scale invariant (GIBOWICZ and KIJKO, 1994).

This note presents preliminary results from cluster analyses of mining-induced seismicity in a deep South African gold mine. For some time now it has been speculated that clusters of seismicity in mines have a definite effect on one another even over long distances. In this paper the interaction of seismicity using cross-correlation of time histories of activity rates and radiated energy between all pairs of clusters is quantified. The degree of interaction as a function of distance can be determined by the correlation coefficients between the groups of events.

Space-time Clusters of Seismicity

The formulation of the clustering process used herein is close to the single-link cluster analysis first introduced in seismology by FROHLICH and DAVIS (1990) and DAVIS and FROHLICH (1991). The difference in the approach used here is the incorporation of a moving time window which is similar to the procedure used by MATSUMURA (1984). For each event within the window, links are calculated between the most recent event and all other events in the window. The link distance is defined in both a space and time as

$$d_{\mathrm{ST}} = \sqrt{d^2 + c^2 t^2} \tag{1}$$

where d is the space distance between events, t the time difference and c a time-distance conversion factor to be defined later. If a link is found to be less than the certain specified maximum link length $\mathrm{MAX}(d_{\mathrm{ST}})$, the two events in question are linked together. At some stage two clusters could physically be linked together, however no events occurred near enough to each other for such a link to be formed. This situation can arise from using a time window which is too small. Such a problem is easily amended by periodically testing distances between cluster centroids. If cluster centroids drift near enough to each other, the algorithm will merge them into one.

The time window length is therefore one of the most important parameters. If the selected window length is too small, clusters may have no events in the window at some point in time. Such a situation would effectively terminate the cluster. If the window length is too large, many links must be calculated for each new event, causing the algorithm to become unnecessarily slow. The average activity rate for the mine must be studied to decide on an appropriate window length. From our experience, the window should have at least 25 events at any time.

In addition to the time window, the method requires the specification of two parameters, the "cutoff distance" $MAX(d_{ST})$, and the time-distance conversion factor c. A value of c is chosen such that two simultaneous events separated by a distance d are as likely to be related to each other as two events with identical locations occurring at time t apart. In practice, c factor is determined as a mean value of d/t ratio for all events in the catalog.

The ability of the space-time clustering analysis algorithm to detect anomalous clusters of seismicity is demonstrated in Figure 1. Data used in the analysis cover the time period of one year of seismic activity at one gold mine in South Africa. In total 20,024 seismic events were analyzed with seismic energy equal to or greater

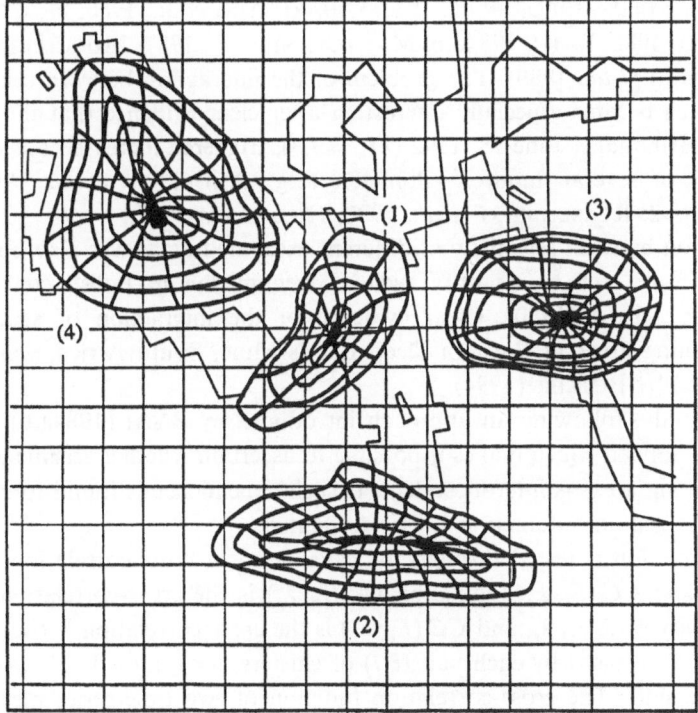

Figure 1
Four from seven identified clusters of seismic events at analyzed gold mine in South Africa.

than 10^4 J. A moving time window of 10 days was used with a maximum space-time distance $\text{MAX}(d_{ST})$ equal to 250 m. Without incorporating any knowledge of the mining, the space-time clustering algorithm was able to detect seven clusters of seismicity in the mine. Four of the clusters and some details of them are demonstrated on Figure 1. The maximum distance between "centers" of the clusters did not exceed 1800 m.

Correlation Between Clusters

An interesting result from the detailed analysis of sesimicity at Western Deep Levels Mine, South Africa, was the demonstration of similarities between patterns of probabilities of strong seismic event occurrence within separate clusters (KIJKO et al., 1993). Observed similarities suggest the presence of mutual interactions between seismic event occurrence at different clusters. Unfortunately, in the quoted work, no effort was made to test and quantify observed interactions.

The concept of interaction between natural seismicity at different areas is not new. This problem was studied by several authors, and different approaches are known (e.g., KEILIS-BOROK et al., 1972; CHIARUTTINI et al., 1980; PROZOROV and DZIEWONSKI, 1982; BÅTH, 1984a,b; MANTOVANI et al., 1987; MUCCIARELLI et al., 1988; ALBERTO et al., 1989). The problem of the interaction between seismicity at different areas became especially interesting after clear evidence that the Landers, California earthquake, June 28, 1992 ($M_w = 7.3$), triggered increased seismicity out to distances of several hundred kilometers (e.g., ANDERSON et al., 1992, 1993; MITCHEL, 1992; REASENBERG et al., 1992; HILL et al., 1993).

Interaction between seismicity at different excavated areas was also observed in mines. A simple trial to quantify such a phenomenon in Polish coal mines is described by KIJKO (1980). Clear evidence of the interaction of seismicity of different mining areas at Western Deep Levels Mine, South Africa, was recently demonstrated by BUTLER (1994).

In our study, following an approach introduced by BÅTH (1984a,b), a simple technique is applied which makes it possible to ascertain whether seismic activity in different mining areas is interdependent. If such dependence is found to exist, then the degree of mutual interaction is also estimated.

The interaction is tested through the calculation of cross-correlation functions $CC_\tau(\lambda_i, \lambda_j)$ and $CC_\tau(E_i, E_j)$. Function $CC_\tau(\lambda_i, \lambda_j)$ is the cross-correlation between seismic activity rates λ_i, λ_j, and $CC_\tau(E_i, E_j)$ is the cross-correlation between seismic energies E_i, E_j released by each pair (i, j) of clusters, $i < j$, $i, j = 1, \ldots, 4$, and τ is mutual time shift. The cross-correlation function of two time series measures the similarity between them as a function of a relative time shift. The normalized cross-correlation function of two discrete time series $\{x\} = \{x_0, \ldots, x_n\}$ and

$\{y\} = \{y_0, \ldots, y_n\}$ is defined as (e.g., DIMRI, 1992)

$$CC_\tau(x, y) = \frac{1}{(n-s)\,\sigma_x\,\sigma_y} \sum_{k=0}^{n-s} (x_k - \bar{x})(y_{k+s} - \bar{y}), \qquad (2)$$

where time shift $\tau = s \cdot \Delta t$, Δt is sampling rate, s is an integer, $\bar{x}$, $\bar{y}$, and σ_x, σ_y are averages and standard deviations of time series $\{x\}$ and $\{y\}$, respectively. The value of CC_τ is always between $\langle -1, 1 \rangle$. If it is $+1$, the time series $\{x\}$ and $\{y\}$ have a perfect positive linear relationship. If CC_τ is equal to -1, $\{x\}$ and $\{y\}$ are linearly related but in the opposite way. If there exists no relationship between $\{x\}$ and $\{y\}$, then the value of CC_τ is 0. Intermediate values indicate partial relationships

An assessment of the interaction between four detected clusters of seismicity is presented below. Before calculating cross-correlation functions, the raw data were cumulated into $\Delta t = 3$ week time intervals (Table 1). Visual scanning reveals a high resemblance between the curves displayed in the set of Figure 2. Not only the major features, but also several synchronous details can be seen in each set of respective time series.

In our analysis, the cross-correlation functions of activity rate and logarithm of released seismic energy between each pair of four clusters was calculated. Thereafter the maximum of the absolute value of cross-correlation functions was used as an

Table 1

Number of seismic events and seismic energy released in 3-week time intervals, for 4 clusters for the time period of one year

Time interval (in 3-weeks units)	Nos. of events				log energy [J]			
	Cluster nos.				Cluster nos.			
	1	2	3	4	1	2	3	4
1	24	4	9	4	157.7	26.9	54.5	23.8
2	12	1	7	8	79.2	5.9	47.6	51.1
3	12	1	12	6	71.7	4.9	74.3	39.2
4	14	8	10	5	88.4	51.4	56.8	29.9
5	21	7	15	6	122.7	38.8	78.1	33.9
6	21	7	21	5	129.2	46.1	123.9	29.5
7	20	3	7	7	129.6	16.2	37.5	42.3
8	14	3	9	2	90.2	18.9	53.7	15.4
9	25	11	19	7	145.5	64.2	108.5	45.2
10	29	4	18	13	177.3	19.7	102.6	75.9
11	38	15	29	12	217.7	78.6	156.1	68.2
12	41	15	33	17	230.2	77.6	174.7	86.7
13	23	5	14	13	148.9	32.2	86.8	83.3
14	17	7	15	16	104.9	36.9	88.2	90.6
15	35	13	25	37	190.4	70.7	143.4	188.9
16	72	13	40	33	403.1	74.1	230.4	153.2
17	49	18	36	34	256.7	95.5	194.1	173.8

Table 2

Maximum cross-correlation coefficients of activity rates (upper triangular matrix) and seismic energy released (lower triangular) for 6 possible pairs of 4 clusters

		Activity rate correlation			
		1	2	3	4
Seismic energy release correlation	1		0.82	0.89	0.77
	2	0.79		0.87	0.64
	3	0.91	0.88		0.75
	4	0.79	0.69	0.77	

indicator of interaction. Maxima of cross-correlations of seismic activities, reach values from 0.65 to 0.89 and respective cross-correlations of seismic energy release vary between 0.69 and 0.91. The maxima of cross-correlation functions are summarized in Table 2. With only one exception (clusters # 1 and # 4), the maxima occur at $\tau = 0$ time lag. It is interesting to note that when the data are cumulated into one week intervals, the maxima of the cross-correlation functions remain close to $\tau = 0$.

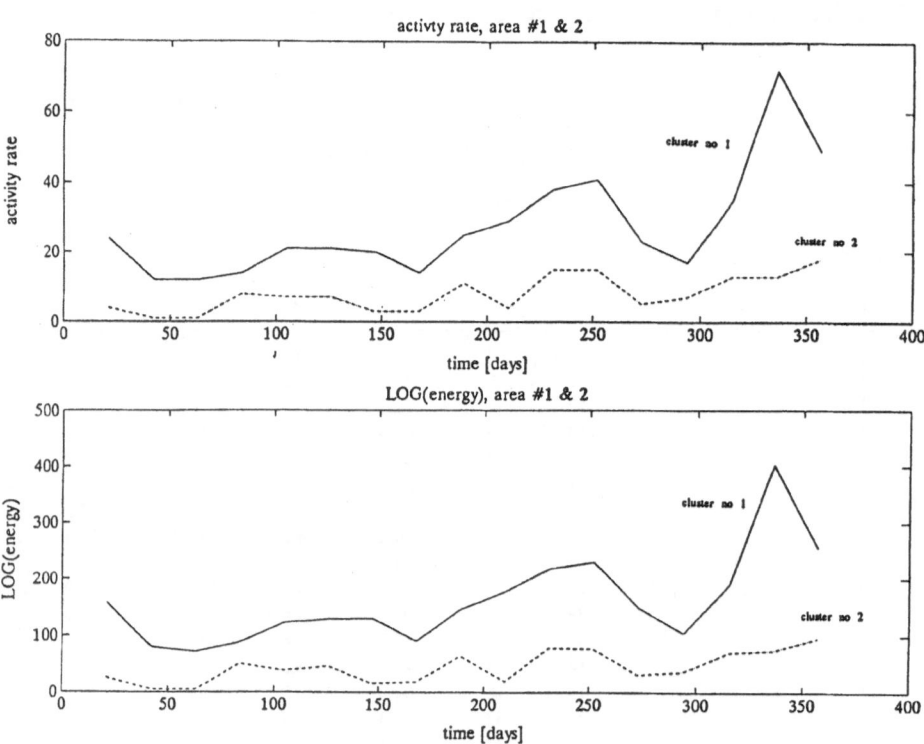

Figure 2(a)

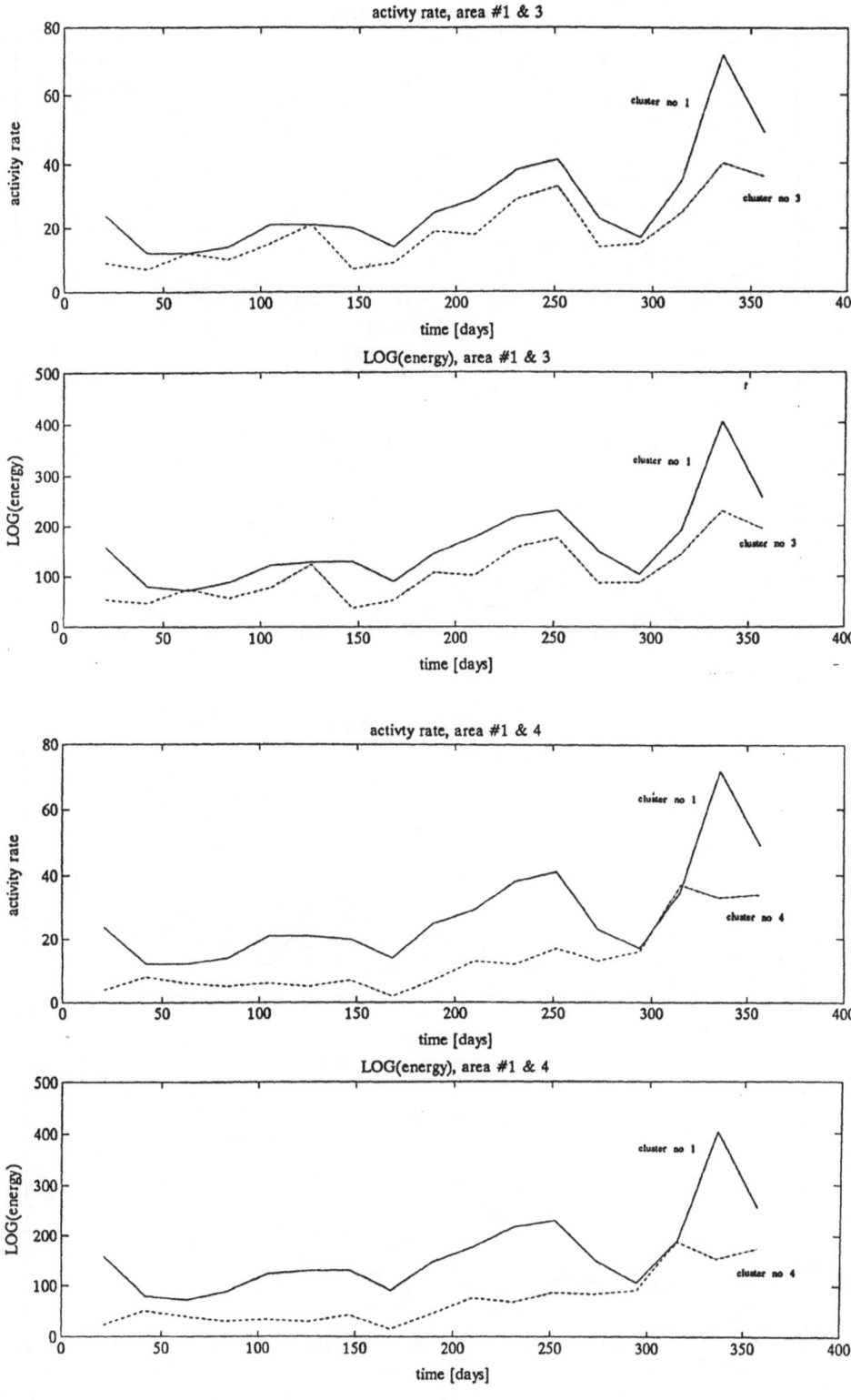

Figures 2(b), (c)

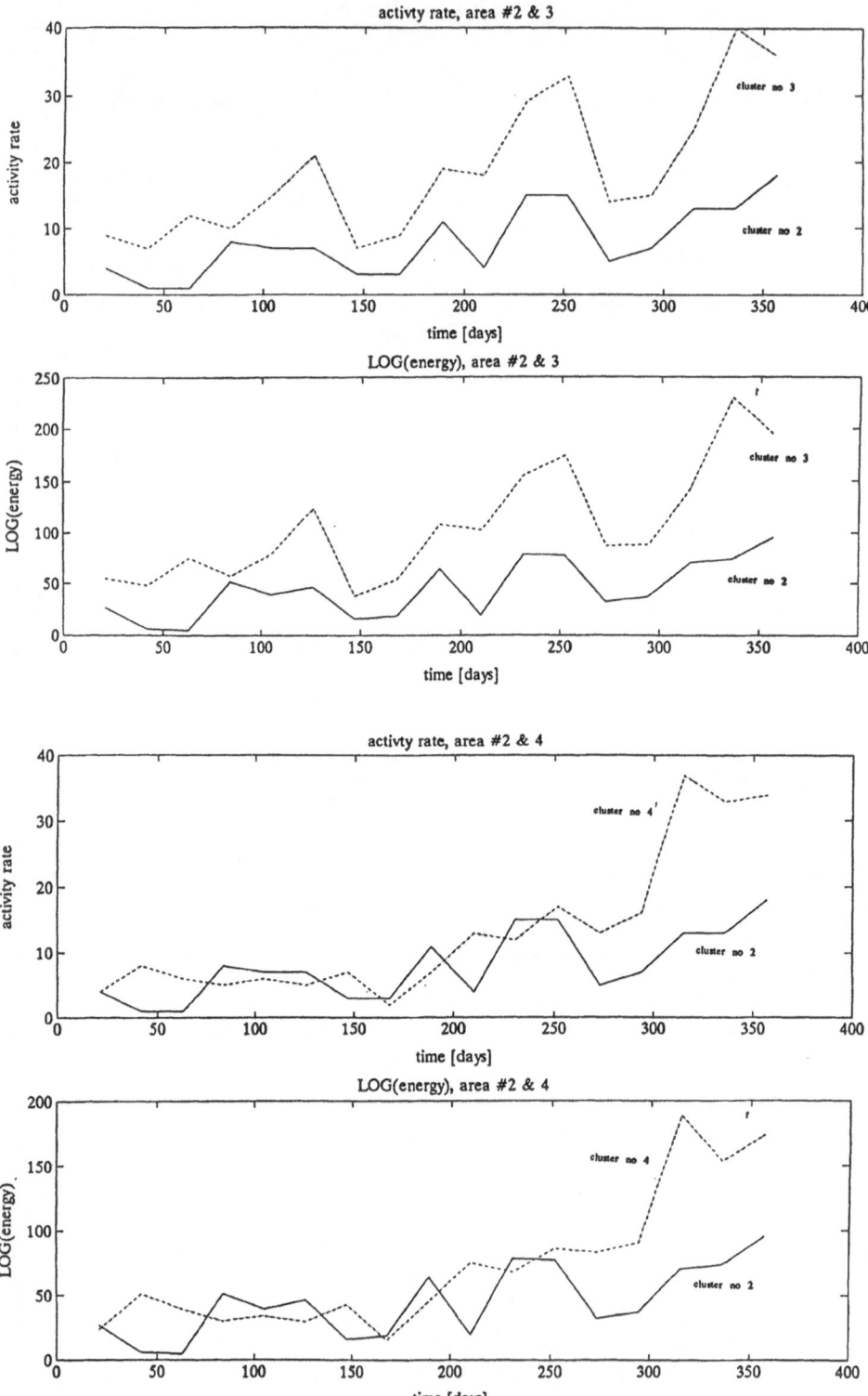

Figures 2(d), (e)

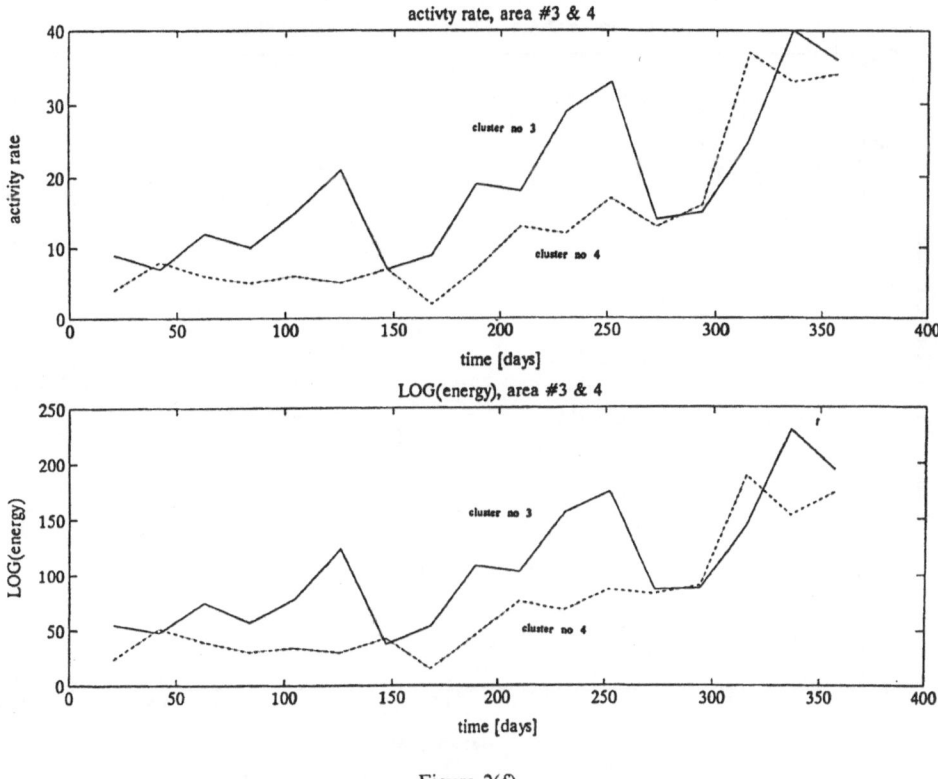

Figure 2(f)

Figure 2
Number of seismic events (upper plot) and logarithm of seismic energy released (lower plot) in 3-week time intervals, for 6 possible pairs of 4 clusters for the time period of one year. (a) clusters 1 and 2; (b) clusters 1 and 3; (c) clusters 1 and 4; (d) clusters 2 and 3; (e) clusters 2 and 4; (f) clusters 3 and 4.

An unexpected result is that the simple technique, used for evaluation of cluster interaction, would also manifest a decrease of interaction with the distance between clusters. This is in fact the case (Fig. 3). Correlation coefficients for six distances are plotted in this figure as a function of distance from a reference cluster. This result shows decreasing interaction and is encouraging, since intuitively one would expect this.

The results of the performed analysis may be summarized as follows: (1) there is a clear correlation between seismic activity and seismic energy release in detected clusters; (2) in most cases, time shifts associated with correlation maxima are less than 21 days; (3) the degree of interaction between clusters decreases as distance increases.

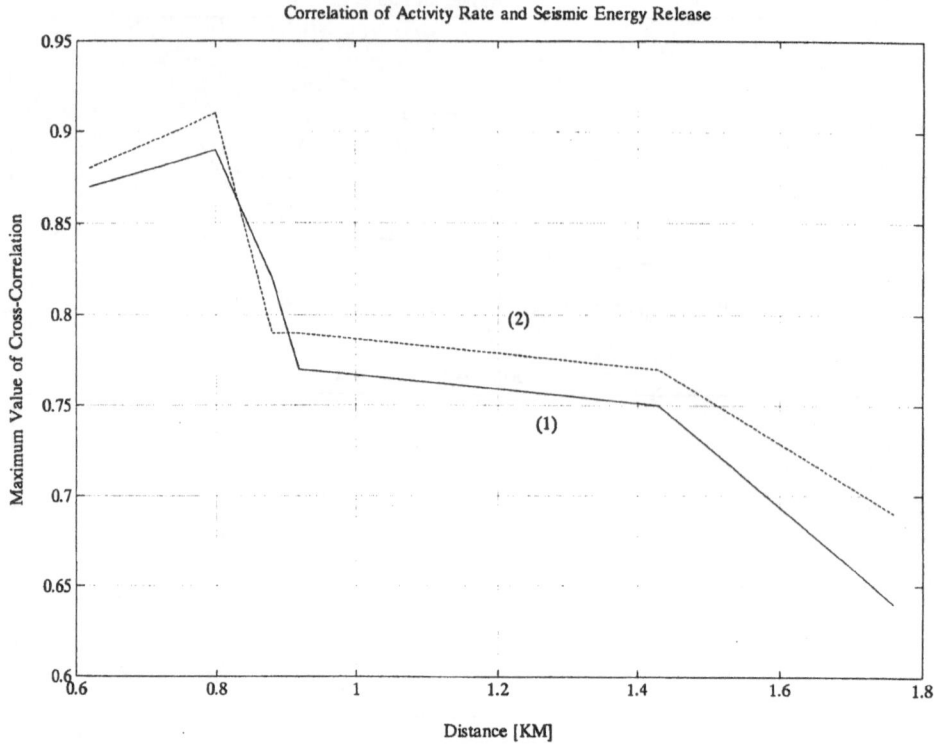

Figure 3
Correlation coefficients of activity rates (1) and released seismic energy (2) as a function of distance from
a reference cluster.

Conclusions

Mining engineers have suspected for some time now that seismicity at active faces is affected by induced seismicity at other mining faces, in some cases at significant distances apart. The results from this study support this and further indicate that the degree of interaction is indeed a function of distance. One must therefore consider fairly large regions of activity around the zone of interest when performing back or forward analyses. The result also suggests that triggering mechanisms may be more complex than was previously suspected.

Acknowledgement

Numerous comments and suggestions by T. Hagan and M. Sciocatti greatly improved an early version of this article.

This work was partially supported by the Department of Mineral and Energy Affairs of South Africa, as a project entitled: *Strategies, Methodologies and Technologies for Seismic Monitoring, Analysis and Interpretation in Rockbursts Prone Mines.*

REFERENCES

ALBERTO D., MUCCIARELLI, M., and MANTOVANI, E. (1989), *Use of Non-parametric Correlation Tests for the Study of Seismic Interrelations*, Geophys. J. *96*, 185–188.

ANDERSON, J. G., LOUIE, J., BRUNE, J. N., DE POLO, D., SAVAGE, M., and YU, G. (1992), *Seismicity in Nevada Apparently Triggered by the Landers, California, Earthquake, June 28, 1992* (abstract), EOS Supplement, 1992 Fall Meeting, American Geophysical Union, 393.

ANDERSON, J. G., BRUNE, J. N., LOUIE, J., ZENG, Y., SAVAGE, M., YU, G., CHEN, Q., and DE POLO, D. (1993), *Seismicity in the Western Great Basin Apparently Triggered by the Landers, California Earthquake, June 28, 1992*, Preprint, August 3, 1993, 19 pp.

BÅTH, M. (1984a), *Correlation between Regional and Global Seismic Activity*, Tectonophysics *104*, 187–194.

BÅTH, M. (1984b), *Correlation between Greek and Global Seismic Activity*, Tectonophysics *109*, 345–351.

BUTLER, A. G. (1994). Unpublished manuscript.

CHIARUTTINI, C., KIJKO, A., and TEISSEYRE, R. (1980), *Tectonic Discrimination of the Fruli Earthquakes*, Bull. Geofis. Ther. Appl. *22*, 295–302.

DAVIS, S. D., and FROHLICH, C. (1991), *Single-link Cluster Analysis, Synthetic Earthquake Catalogs, and Aftershocks Identification*, Geophys. J. Int. *104*, 289–306.

DIMRI, V., *Deconvolution and Inverse Theory. Application to Geophysical Problems* (Elsevier, Amsterdam 1992).

DZIEWONSKI, A. M., and PROZOROV, A. G. (1984), *Self-similar Determination of Earthquake Clustering*, Comp. Seism. *16*, 7–16.

ENEVA, M., and PAVLIS, G. L. (1988), *Application of Pair Analysis Statistics to Aftershocks of the 1984 Morgan Hill, California Earthquake*, J. Geophys. Res. *93*, 9113–9125.

FROHLICH, C., and DAVIS, S. D. (1990), *Single-link Cluster Analysis as a Method to Evaluate Spatial and Temporal Properties of Earthquake Catalogs*, Geophys. J. Int. *100*, 19–32.

GIBOWICZ, S. J., and KIJKO, A., *An Introduction to Mining Seismology* (Academic Press, San Diego 1994).

HABERMANN, R. E., (1988), *Precursory Seismic Quiescence: Past, Present and Future*, Pure and Appl. Geophys. *126*, 279–318.

HILL, D. P., REASENBERG, P. A., MICHAEL, A., ARABASZ, W. J., BEROZA, G., BRAUMBAUGH, D., BRUNE, J. N., CASTRO, R., DAVIS, S., DE POLO, D., ELLSWORTH, W. L., GOMBERG, J., HARMSEN, S., HAUSE, L., JACKSON, S. M., JOHNSTON, M. J. S., JONES, L., KELLER, R., MALONE, S., MAUNGUIA, L., NAVA, S., PECHMANN, J. C., SANFORD, A., SIMPSON, R. W., SMITH, R. W., STARK, M., STICKNEY, M., VIDAL, A., WALTER, S., WONG, V., and ZOLLWEG, J. (1993), *Seismicity Remotely Triggered by the Magnitude 7.3 Landers, California Earthquake*, Science *260*, 1617–1623.

JOHNSON, C., KEILIS-BOROK, V. I., LAMORE, R., and MINISTER, B. (1984), *Swarms, of Main Shocks in Southern California*, Comp. Seism. *16*, 1–6.

KAGAN, Y. Y. (1981), *Spatial Distribution of Earthquakes: Four-point Moment Function*, Phys. Earth Planet. Int. *12*, 291–318.

KAGAN, Y. Y., and JACKSON, D. D. (1991a), *Long-term Earthquake Clustering*, Geophys. J. Int. *104*, 117–133.

KAGAN, Y. Y., and JACKSON, D. D. (1991b), *Seismic Gap Hypothesis: Ten Years After*, J. Geophys. Res. *96*, 21419–21431.

KAGAN, Y. Y., and KNOPOFF, L. (1980), *Spatial Distribution of Earthquakes: The Two Point Correlation Function*, Roy. Astron. Soc. *62*, 303–320.

KEILIS-BOROK, V. I., PODGAETSKAYA, V. M., and PROZOROV, A. G., *Local statistics of earthquake catalogs*. In *Computational Seismology* (ed. V. I. Keilis-Borok) (Consultants Bureau, New York 1972) pp. 214–227.

KIJKO, A. (1980), *Statistical Test of Mutual Dependence of Seismic Activities in Two Adjacent Regions*, Publ. Inst. Geophys. Pol. Acad. Sci. *A-10* (142), 125–133.

KIJKO, A., FUNK, C. W., and BRINK, A. V. Z. (1993), *Identification of anomalous patterns in time-dependent mine seismicity*. Proceedings of the 3rd International Symposium on Rockbursts and Seismicity in Mines, 16–18 August 1993, Kingston, Ontario, Canada, 205–210.

LEI, X., NISHIZAWA, O., and KUSUNOSE, K. (1993), *Band-limited Heterogeneous Fractal Structure of Earthquakes and Acoustic-emission Events*, Geophys. J. Int. *115*, 79–84.

MANDELBROT, B. B., *The Fractal Geometry of Nature* (Freeman, San Francisco 1982).

MANDELBROT, B. B. (1989), *Multifractal Measures, Especially for the Geophysicists*, Pure and Appl. Geophys. *131*, 5–42.

MANTOVANI, E., MUCCIARELLI, M., and ALBERTO, D. (1987), *Evidence of Interrelation between the Seismicity of the Southern Apennines and Southern Dinarides*, Phys. Earth Planet. Interiors *49*, 259–263.

MATSUMURA, S. (1984), *A One-parameter Expression of Seismicity Patterns in Space and Time*, Bull. Seismol. Soc. Am. *74*, 2529–2576.

MICHAEL, A. W. (1992), *Initiation of Seismicity Remotely Triggered by the Landers Earthquake: Where and When* (abstract), EOS Supplement, 1992 Fall Meeting, American Geophysical Union, 392–393.

MUCCIARELLI, M., ALBERTO, D., and MANTOVANI, E. (1988), *Earthquake Forecasting in Southern Italy on the Basis of Logistic Models*, Tectonophysics *152*, 153–155.

NATALE, G. D., and ZOLLO, A. (1986), *Statistical Analysis and Clustering Features of the Phlegraean Fields Earthquake Sequence*, Bull. Seismol. Soc. Am. *76*, 801–814.

PROZOROV, A. G., and DZIEWONSKI, A. M. (1982), *A Method of Studying Variations in the Clustering Property of Earthquakes: Application to the Analysis of Global Seismicity*, JGR *87*, 2829–2839.

REASENBERG, P. A., and MATTHEWS, M. V. (1988), *Precursory Seismic Quiescence: A Preliminary Assessment of the Hypothesis*, Pure and Appl. Geophys. *126*, 373–406.

REASENBERG, P. A., HILL, D. P., MICHAEL, A. J., SIMPSON, R. W., ELLSWORTH, W. L., WALTER, S., JOHNSTON, M., SMITH, R., NAVA, S. J., ARABASZ, W. J., PECHMANN, J. C., GOMBERG, J., BRUNE, J. N., DE POLO, D., BEROZA, G., DAVIS, S. D., and ZOLLWEG, J. (1992), *Remote Seismicity Triggered by the M7.5 Landers, California Earthquake of June 28, 1992* (abstract), EOS Supplement, 1992, 1992 Fall Meeting American Geophysical Union, 392.

SMALLEY, R., Jr., CHATELAIN, J.-L., TURCOTTE, D., and PRÉVOT, R. (1987), *A Fractal Approach to the Clustering of Earthquakes: Applications to the Seismicity of the New Hebrides*, Bull. Seismol. Soc. Am. *77*, 1368–1381.

WARDLAW, R. L., FROHLICH, C., and DAVIS, S. D. (1990), *Evaluation of Precursory Seismic Quiescence in Sixteen Subduction Zones Using Single-link Cluster Analysis*, Pure and Appl. Geophys. *134*, 57–78.

WYSS, M., and HABERMANN, R. E. (1988), *Precursory Seismic Quiescence*, Pure and Appl. Geophys. *126*, 319–332.

XIE, H., and PARISEAU, W. G. (1993), *Fractal Character and Mechanism of Rock Bursts*, Int. J. Rock Mech. Min. Sci. and Geomech. Abstr. *30*, 343–350.

(Received January 5, 1995, revised August 17, 1995, accepted August 24, 1995)

PAGEOPH, Vol. 147, No. 2 (1996)

0033–4553/96/020289–16$1.50 + 0.20/0

Rock-mass Characterization Using Intrinsic and Scattering Attenuation Estimates at Frequencies from 400 to 1600 Hz

ANDREW J. FEUSTEL,[1,3] CEZAR-IOAN TRIFU[2] and THEODORE I. URBANCIC[2]

Abstract—Intrinsic and scattering S-wave quality factors (Q_β) were estimated using the Multiple Lapse Time Window Analysis (MLTWA) for microseismic events ($M < -1$) with source-sensor distances of 45 to 120 m, associated with an excavation at 630 m depth in Strathcona Mine, Sudbury, Canada. Additional information on the rock mass was provided by underground structural mapping data. Intrinsic Q_β values, at 800 Hz, were on the order of 140, similar to quality factor values obtained in previous studies using Spectral Decay and Coda-Q methods (120 to 170). The scattering quality factor at this frequency was about 520. An observed frequency dependence of the scattering attenuation suggested that a decrease in the density of scatterers, with scale lengths on the order of 2 m, exists at the site. Characteristic fracture scale lengths were considered to range from 4 to 6 m as identified in the mapping data. These observations were supported by the increase in scattering found for seismic waves with frequencies less than 1000 Hz. By assuming that the identified scatters are characteristic faults, these scatterers can then be considered to increase nonsimilar behavior in source scaling. Overall, our results suggest that MLTWA provides a practical method for remotely characterizing the quality of a rock mass when visual observations are not attainable.

Key words: Intrinsic and scattering attenuation, multiple lapse time window analysis, characteristic fracture scale lengths, seismic source scaling.

Introduction

In the past decade, applied seismology has become increasingly important in the operation and development of underground mines. Scaling laws and specialized seismic monitoring equipment have made it possible to apply regional and teleseismic methods to underground mines where study areas, path lengths, and source volumes are several orders of magnitude smaller in dimension. Since seismic source studies rely on the measurement of both time and spectral amplitudes of the observed signals, it is important that the seismic energy loss be properly taken into account. This is especially important for large induced seismic events, where strong ground motions associated with radiated seismic energy pose both safety and mine stability concerns. If path attenuation and rock quality can be established in close

---

[1] Department of Geological Sciences, Queen's University, Kingston, Canada, K7L 3N6.
[2] Engineering Seismology Group Canada, Kingston, Canada, K7L 2Z4.
[3] Now at Engineering Sesimology Group Canada, Kingston, Canada, K7L 2Z4.

proximity to excavations, more accurate peak particle velocities can be estimated and used for ground support development. For example, a low attenuation (high quality factor Q) may be indicative of a tightly bound rock mass, or one that is capable of propagating a displacement pulse with little or no energy loss, thus reaching the free surface with nearly full strength.

Various techniques have been proposed in earthquake seismology to evaluate attenuation effects, such as the Displacement Spectral Decay (DSD) method, Rise-Time method, empirical Green's function (eGf), Coda-Q method, and the Multiple Lapse Time Window Analysis (MLTWA). The use of each technique requires that specific criteria be met in terms of source function assumptions, sensor array geometry, signal quality and attenuation mechanisms. The application of these techniques has been almost exclusively used to investigate regional and global seismic attenuation (e.g., SCHERBAUM, 1990; MAYEDA *et al.*, 1991). However, few studies have investigated Q at the scale of the rock mass as encountered in underground mining. For example, changes in Coda-Q in the vicinity of an underground opening, assuming a single scattering model, were studied by CICHOWICZ and GREEN (1989) and CICHOWICZ *et al.* (1990). Their studies reported P- and S-wave Q values ranging from 20 to 100 next to the opening, and up to 300 outside the region of excavation influence. Using a DSD approach, SPOTTISWOODE (1993) found P- and S-wave Q values between 20, for ray-paths traversing excavation areas, and 1000, for ray paths in competent rock. Using a similar technique, FEUSTEL *et al.* (1993) estimated Q values of approximately 120 for both P and S waves generated by events of magnitude less than -1 associated with an excavation at depth. FEUSTEL and YOUNG (1994) estimated S-wave Q values from Spectral Ratios and MLTWA for artificial sources (between 2 and 5 kHz) at an underground research laboratory. They found intrinsic Q values of 85 to 115 and scattering Q values of greater than 800 for source-sensor distances of 10 to 80 m.

In this study, we examine intrinsic attenuation and scattering estimates as obtained through the use of MLTWA for frequencies ranging from 400 to 1600 Hz. The advantages of using this technique over other methods are its source independence, ability to analyze passively acquired seismicity, and use of recently developed scattered wave energy models. Our analysis considers microseismic events, with magnitudes < -1, recorded with an underground seismic network surrounding an excavation at 630 m depth in Strathcona mine, Sudbury, Canada. Interpretation of the results is supplemented with underground structural mapping data compiled for the region. Based on this comparison, we discuss the appropriateness of the MLTWA approach for investigating characteristic attenuation mechanisms at the scale of the rock mass.

Geomechanical Setting and Microseismic Data

Strathcona Mine is an active nickel-copper mine located on the north rim of the Sudbury Basin, Ontario (Fig. 1). The ore, which occurs in a sulphide-rich breccia

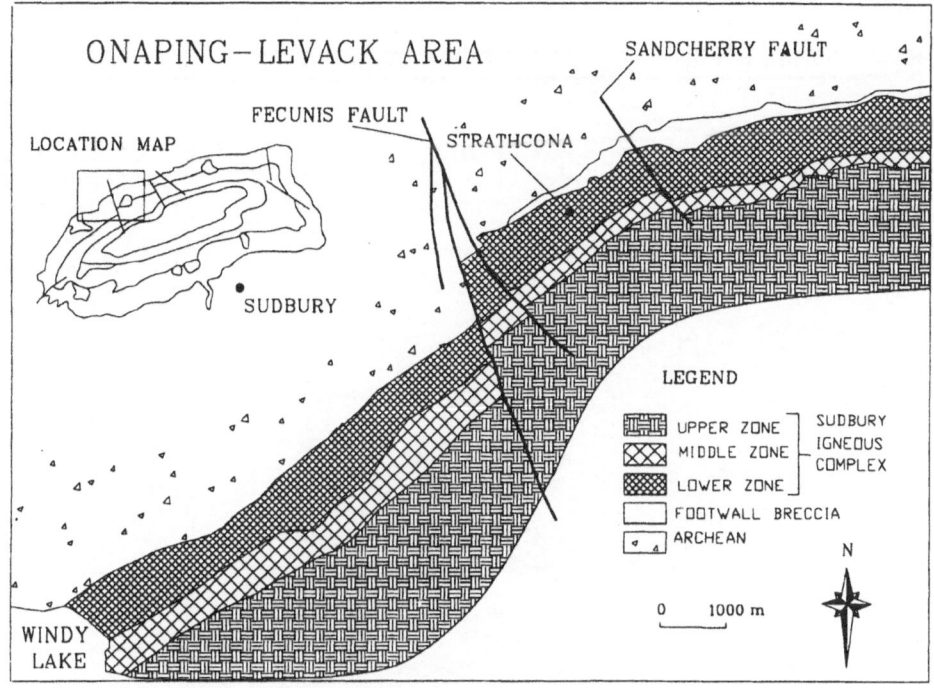

Figure 1
Regional setting of Strathcona mine relative to the Sudbury Basin.

unit, lies between mafic norites (hanging wall) and a granitic-gneiss complex (footwall). The footwall unit is composed of fine to coarse grained hypersthene quartz diorite (COATS and SNAJDR, 1984). In this study, the attentuation effects are considered for seismic events occurring predominantly within the footwall and associated with an excavation ranging in depth from 605 to 640 m. Underground mapping of fractures in this region indicated that the fracturing is dominated by a single structural set (Fig. 2a), trending east-west and moderately dipping to the north (BIRD, 1993). A weakly defined north-south sub-vertical fracture set is also present. Spacing of fractures ranged from 0.3 to 2.6 m, with the highest concentration occurring at about 0.8 m (Fig. 2b). The fracture sets can be considered as being rough and undulating. In general, the fracture persistence was greater than its observed exposure of 3.5 m, which corresponds to the height of tunnel openings. Based on fracture spacing and the number of recorded fracture terminations, the average persistence was estimated at 4 to 6 m. At the depths of interest, the maximum and intermediate principal stresses are sub-horizontal, trending east-west (27 MPa) and north-south (23 MPa), whereas the minimum principal stress is sub-vertical (16 MPa).

(a)

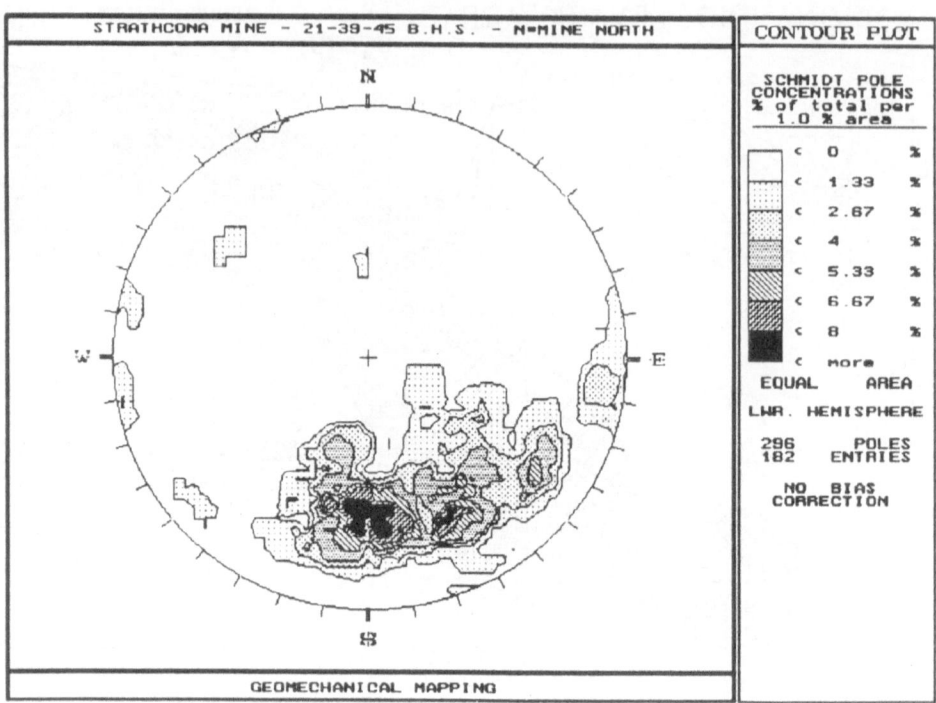

(b)

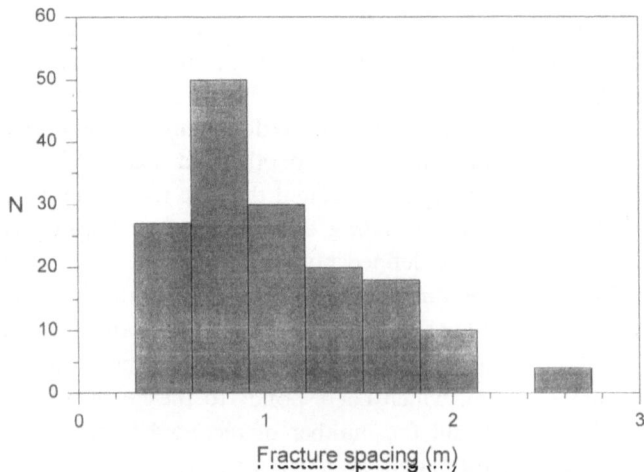

Figure 2

(a) Distributions of poles to fractures in the region of investigation and (b) fracture spacing as determined from underground structural mapping.

An underground 64 channel microseismic system, consisting of five triaxial (dual-gain recording at $\times 3$ and $\times 30$) and 34 uniaxial accelerometers, was installed to monitor the region of interest. The triaxials had a flat response between 1 Hz and 5 kHz and overall sensitivities of 0.9 and 9 V/g, respectively, whereas the uniaxial accelerometers had overall sensitivities of 950 V/g. The sensors were optimally distributed in three dimensions with an inter-sensor spacing of 30 to 60 m. The sampling frequency of the system was set to 20 kHz and the low and high frequency cut-off filters were set to 0.1 and 3.5 kHz, respectively. Seismic data was collected during periods of active mining following production blasts (October, 1991 to January, 1992), as well as during a non-mining period (December, 1991). Locations were obtained for 256 events, using at least 10 manually picked P-wave arrivals ($V_P = 5970$ m/s), to an accuracy of 2 to 3 m. For the attenuation analysis, only triaxial #2 records were used (Fig. 3). The events recorded on this sensor were located within an aperture of about 60° with source-sensor separations of 45 to 120 m, and displayed high signal-to-noise ratios (>4). The ray paths were generally unobstructed by openings and backfill. About 20% of the ray paths, with lengths >100 m, were perturbed by the presence of drifts. This effect was considered minimal as the ray paths were generally affected by a single drift (3 to 5% disturbance over the entire path). The analyzed signals were rotated by an eigen-value/eigenvector decomposition according to the method described by MAT-SUMURA (1981). The method finds the particle vibration orientation of the P wave and rotates the first channel of the triaxial into this ray-trajectory. The second and third channels represent maximum orthogonal particle vibrations to the P wave, corresponding to the S-wave components ($V_S = 3700$ m/s).

Method of Analysis

The MLTWA (FEHLER et al., 1992) is adapted from WU'S (1985) use of radiative transfer theory to model scattered wave energy. The method, which is source independent, attempts to fit observed data to a forward model of scattered S-wave energy based on specified values of intrinsic and scattering attenuation. In order to apply MLTWA, a range of path lengths is required, and consequently, only average attenuation estimates can be obtained. In this approach, the energy $E_n\,(f, r)$ for each seismic pulse, at center frequency f and propagation distance r, is calculated in 3 successive time windows ($n = 1, 2, 3$) of equal size. The window size should correspond to at least 20 times $1/f_c$, where f_c represents the most commonly observed corner frequency in the data set. The energy values are obtained by multiplying the squared amplitude of the velocity spectra by two and integrating over the specified frequency limits. To account for differences in event magnitude and site response, the energies are normalized to a coda reference window, according to the coda-wave method adopted from AKI and CHOUET

(1975). This window is positioned at a time corresponding to at least twice the
S-wave travel time to the largest source-sensor separation. Once normalized, the
three $E_n\,(f, r)$ values are multiplied by $4\pi r^2$ to correct for body-wave geometrical
spreading and plotted simultaneously as a function of path distances. Following
energy determinations, the data is matched to a set of model curves, also corrected
for geometrical spreading, as generated with a hybrid-single-scattering-diffusion

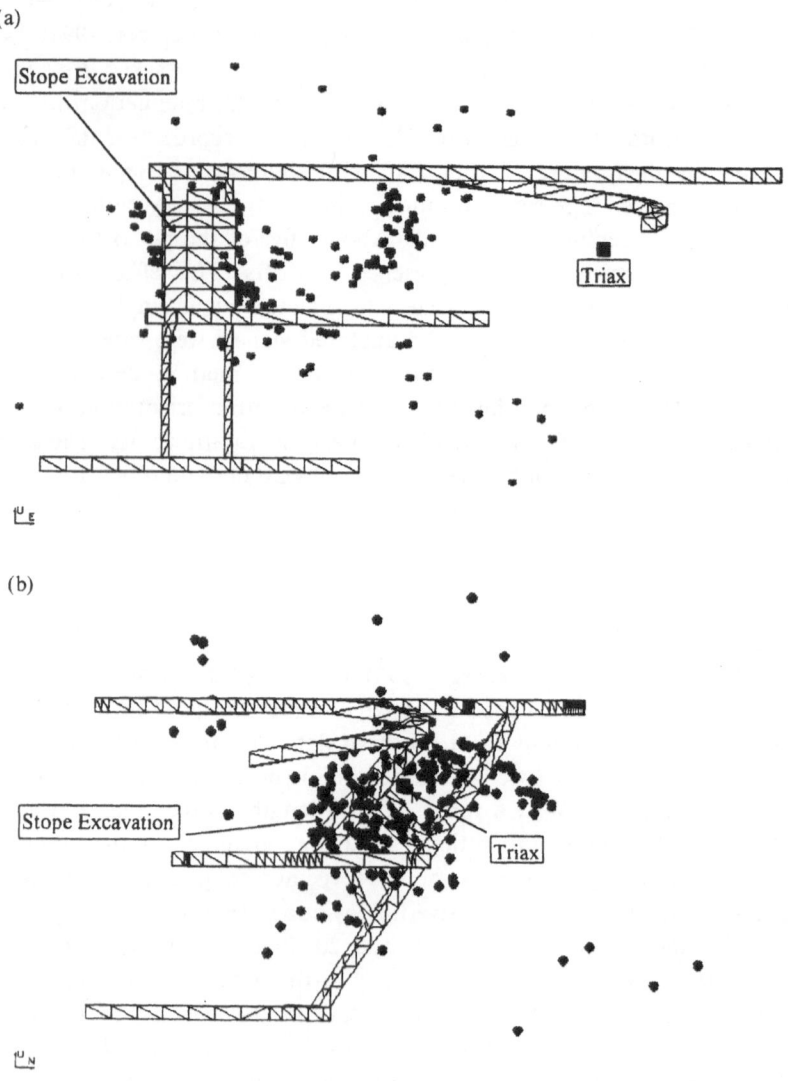

Figure 3

(c)

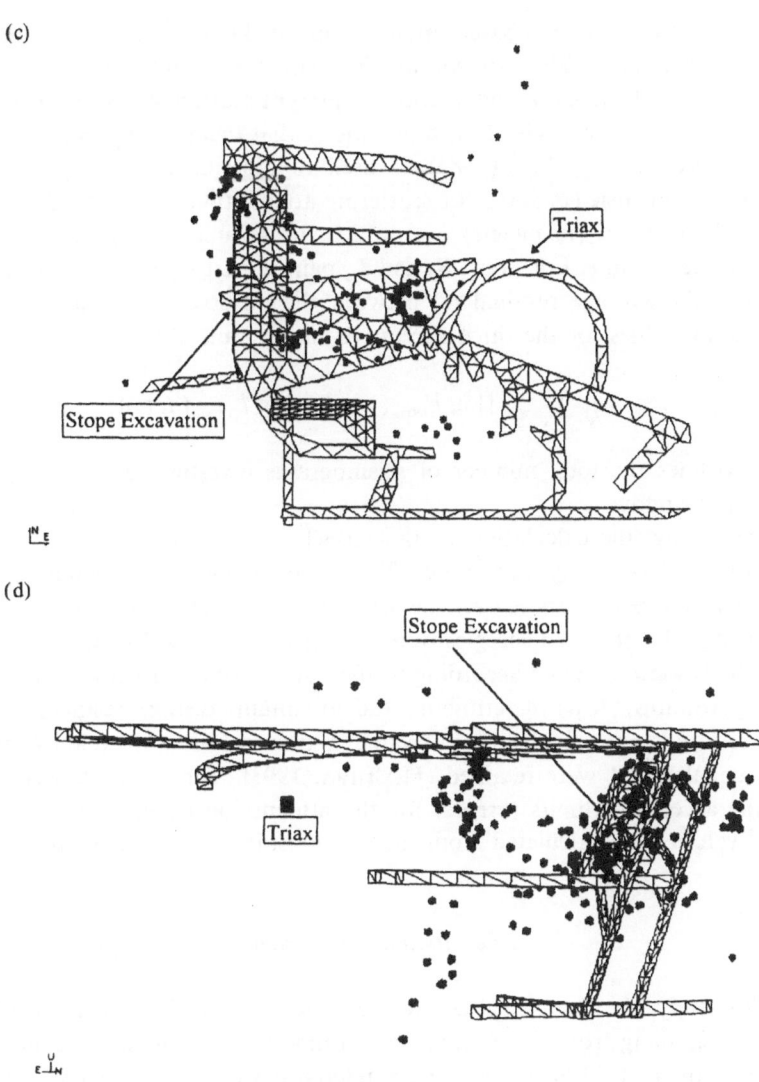

(d)

Figure 3 (Cont.)
Excavation geometry and seismic event locations (circles), relative to triaxial 2 (square), used in the MLTWA looking (a) north, (b) west, (c) down, and (d) in a perspective view to the southwest.

solution for scattered S-wave energy (ZENG, 1991; FEUSTEL, 1995). Each curve represents an approximate solution for the multiple scattered S-wave energy as obtained by combining SATO's (1977) single scattering approximation with a diffusion solution approximation. In this way our technique differs from previous applications of the MLTWA (e.g., HOSHIBA, 1993; JIN *et al.*, 1994) that generated scattered wave energy models with exact solutions from ZENG *et al.* (1991) and HOSHIBA (1991).

Each model is calculated based upon an input value for the seismic albedo (B_0) and total mean path, also known as the extinction length (L_e); where $B_0 = n_s/(n_s + n_i)$, $L_e = 1/(n_s + n_i)$, and n_s and n_i represent scattering and intrinsic attenuation coefficients, respectively. B_0 is a parameter that theoretically ranges from 0 to 1, and, as proposed by WU (1985), is used to describe the energy loss that is dominated by intrinsic ($B_0 < 0.5$) or scattering attenuation ($B_0 > 0.5$). The L_e value describes the distance (in meters) over which the seismic energy becomes e times smaller, where $e = \ln 1$. For every B_0 and L_e pair, a least squares method was used to evaluate the average residual (ε) between the observed (E_{obs}) and theoretical (E_{theo}) energy values for the three successive energy time windows ($n = 1, 2, 3$) as

$$\varepsilon = \frac{1}{N} \sum_{n=1}^{3} \sum_{s=1}^{N} [\log E_{\text{obs}_n}(f, r_s) - \log E_{\text{theo}_n}(f, r_s)]^2$$

where N defines the total number of seismograms investigated or data points in each energy window.

In this study, the calculation of the model curves at every distance was not feasible due to data storage constraints. Therefore, the model values were calculated at 25 equally-spaced distances over the total observed source-sensor distance range. Subsequently, the observed energies were averaged over these 25 adjacent windows. The residuals were weighted according to the number of observed values within the averaging window. After determining the minimum average residual, all other average residuals were divided by this value, and model fits with residual ratios less than or equal to 1.1 were retained (HOSHIBA, 1993). These models were used to define the acceptable limits (errors) for the attenuation coefficients. Correspondingly, Q values were calculated from the relationship $Q_{j=i,s} = 2\pi f/(n_j \beta)$.

Results and Discussion

In this study, the energy values were computed from three successive 22.5 ms windows, beginning from the S-wave arrival time (corresponding to source-sensor separations of 45 to 120 m). The coda reference window (10 ms) was taken at 100 ms from the origin time of each event, corresponding to approximately three times the S-wave travel time for the furthest propagation path length. Three frequency bands, centered at 400 (267–533), 800 (533–1067), and 1600 (1067–2133) Hz were considered for the analysis (Fig. 4). The results for the MLTWA were determined for each frequency band by trial and error. Model curves were calculated for B_0 between 0.1 and 0.9 with a step of 0.04, and L_e between 40 and 140 m with a step of 4 m. After the minimum residual ratio was determined, a supplementary search was carried out around this point by taking half-steps for both B_0 and L_e. The residual ratio results for each frequency band are shown in Figure 5. The results indicate that the attenuation in the monitoring region is

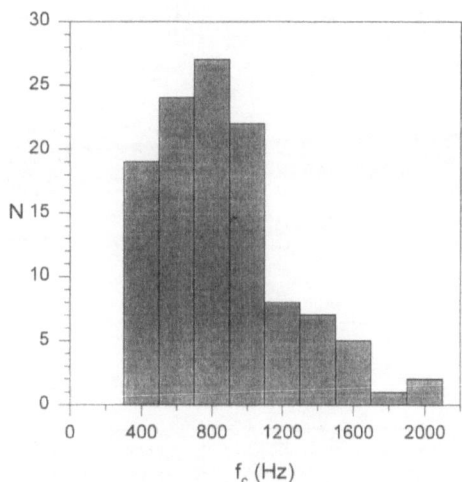

Figure 4
Histogram of the S-wave corner frequencies as determined from the displacement amplitude spectra using the approach outlined by SNOKE (1987).

dominated by intrinsic energy losses ($B_0 < 0.5$), and at 1600 Hz the relative scattering attenuation decreases slightly. At 800 Hz the intrinsic and scattering S-wave quality factors (Q_β) are $146 + 21/-14$, and $519 + 142/-78$, respectively. Figure 6 shows the model fits to the raw data at 800 Hz, as well as model fits to the averaged data. The attenuation coefficients n_s and n_i, corresponding to each of the best-fit Q_β values, are shown as a function of frequency in Figure 7. The error bars represent the range in the coefficients corresponding to model fits with average residuals less than 10% greater than the minimum average residual (determined from the residual ratio analysis; see Fig. 5).

These attenuation results are in agreement with those previously obtained for the study region by employing the DSD approach to a subset of 48 shear mechanism events (FEUSTEL *et al.*, 1993). The spectral decay method relies on BRUNE's (1970) model of the shear source, where all frequencies past the corner frequency have acceleration amplitudes that remain constant, corresponding to a displacement spectral slope of -2. Departures from a ω^{-2} displacement spectral decay are then considered to be caused by attenuation. For comparison purposes, Figure 8 displays the distribution of DSD derived Q estimates, with average values from 120 to 150 ± 40. Compared to the intrinsic attenuation values using MLTWA in the 800 Hz frequency band ($146 + 21/-14$), the results are alike. Similarly, Coda-Q analysis at the site found quality factors of about 170 at a frequency of 800 Hz (FEUSTEL, 1995).

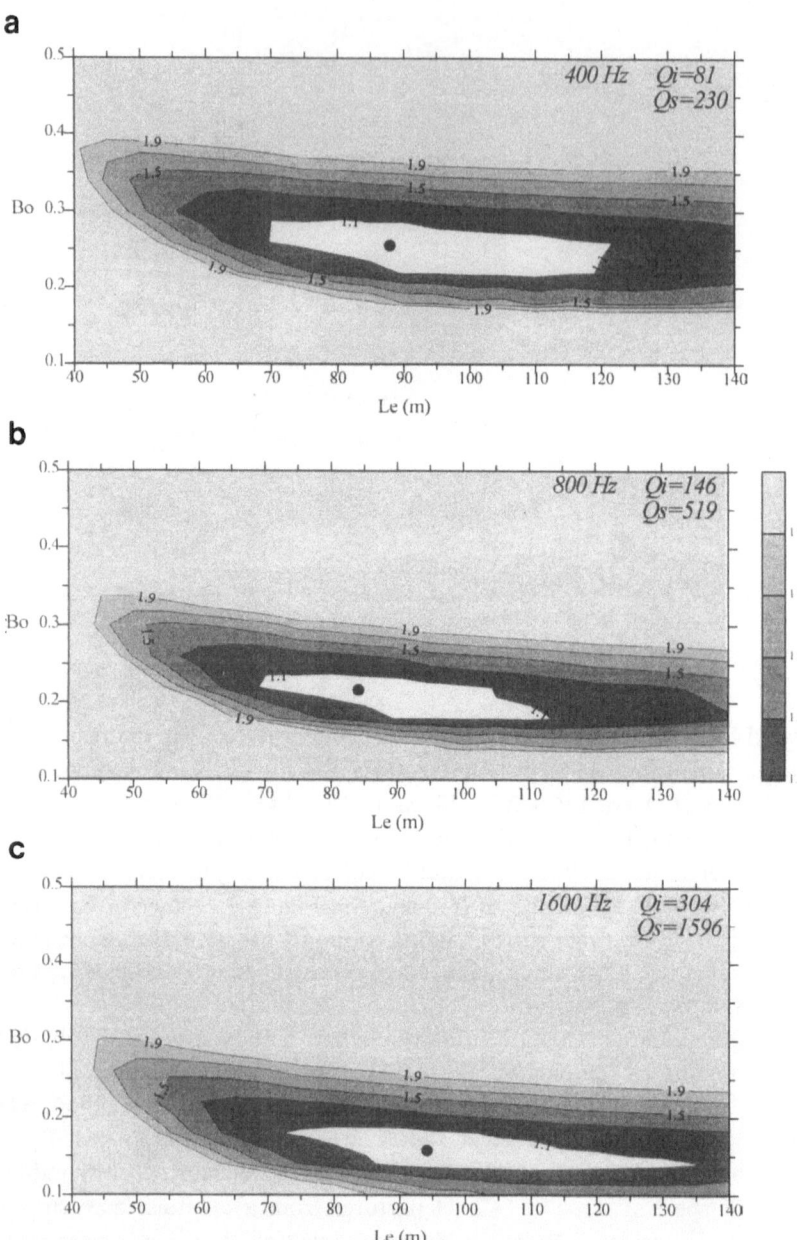

Figure 5
Residual ratio analysis for Strathcona at (a) 400 Hz, (b) 800 Hz, and (c) 1600 Hz. The best fit parameters for seismic albedo (B_0) and total mean free path (L_e) are indicated by a solid circle on the plot. The intrinsic and scattering quality factors corresponding to the best fit models are also given.

(a)

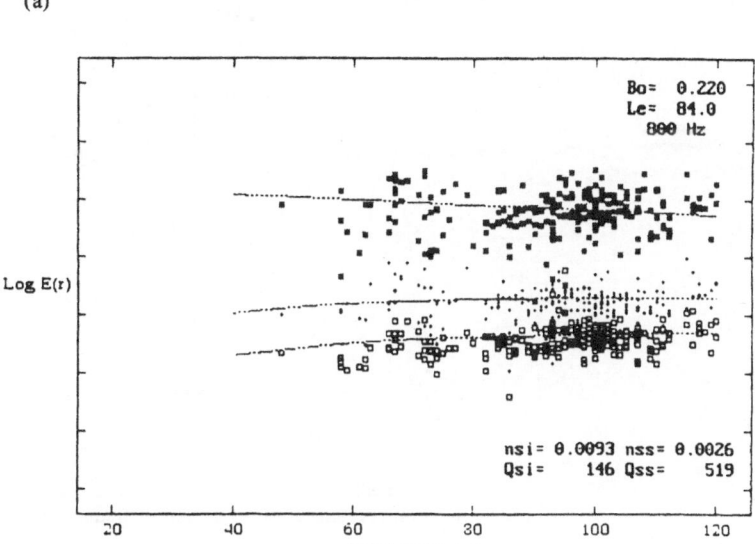

(b)

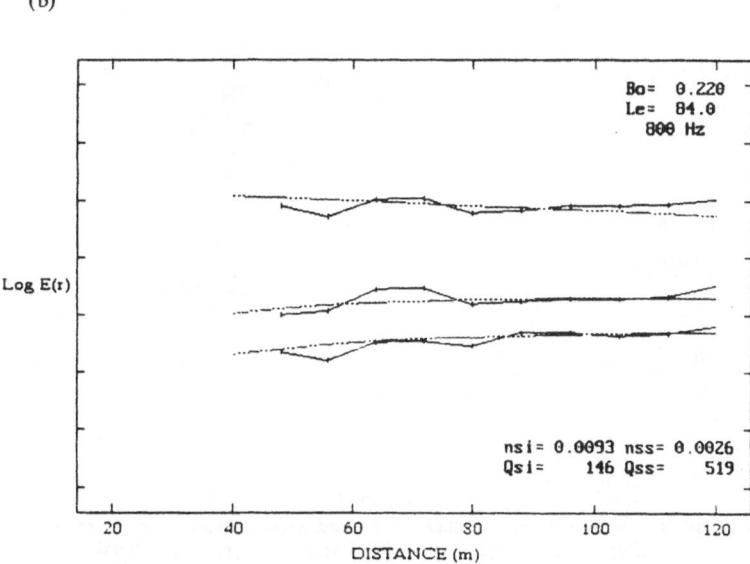

Figure 6
Best fit model (dashed lines) to observed data at 800 Hz. (a) Raw data, where solid squares, dots, and open squares correspond to energy values in three successive time windows. (b) Averaged data, as represented by solid lines with vertical hatch marks. Intrinsic and scattering attenuation coefficients are given along with the intrinsic and scattering quality factors corresponding to the model fit.

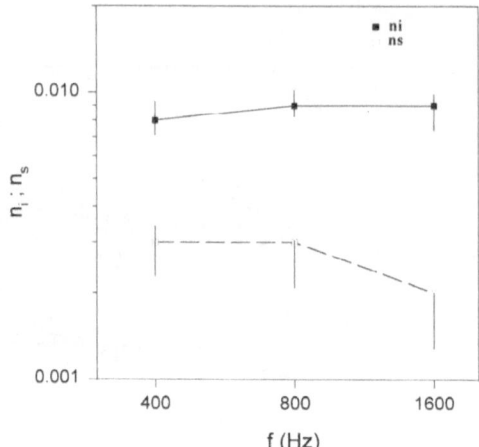

Figure 7
Scattering and intrinsic attenuation coefficients as a function of frequency.

(a)   (b)

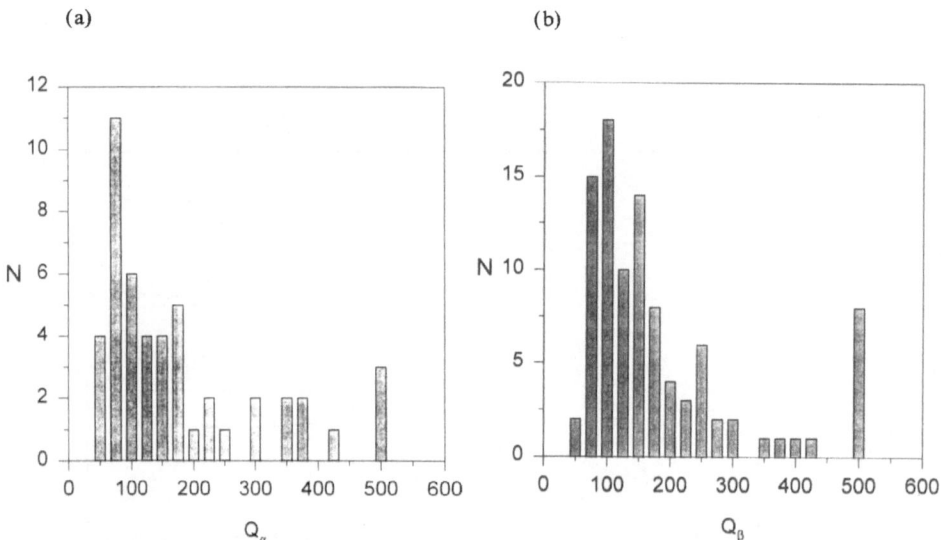

Figure 8
Displacement spectral decay derived estimates of *P*- and *S*-wave quality factors for events occurring within the study volume (modified after FEUSTEL *et al.*, 1993).

From MLTWA, we found that the intrinsic attenuation was not frequency dependent, whereas a frequency dependence was observed for the scattering attenuation. This behavior is believed to be directly related to the average scale-length of heterogeneities in the rock mass. Although the values are relatively constant, between 400 and 800 Hz (Fig. 7), at 1600 Hz the scattering strength decreases, indicating less effect for wavelengths (λ) between 1.75 and 3.5 m. ACHENBACH et al. (1988) indicated that the maximum reflection of a seismic signal occurs when the scale lengths of the heterogeneities in the medium are on the order of the seismic wavelength. Specifically, when the product of the wavenumber ($k = 2\pi/\lambda$) and half-scale length (a) is approximately 1.5 ($0.1 < ka < 10$), backscattering and coda wave generation in the signal are significant. For $ka > 10$, most scattered energy is directed forward, and the backscattered waves are very weak, causing little energy redistribution into the coda waves (WU and AKI, 1988). The average persistence, as determined from underground mapping of fractures, was estimated to range from about 4 to 6 m. At 400 and 800 Hz, the resulting ka values are roughly 2 and 4. These values are within the range of large angle scattering and coda wave generation. However, at 1600 Hz, the ka value increases to approximately 8 and consequently the relative backscattering decreases.

Two additional factors influencing the scattering attenuation were considered: 1) the effect of the relative path and fracture orientations, and 2) the ratio between the dominant fracture spacing and half-fracture persistence. The first is not believed to play a significant role in our observations, due to a 90° aperture in fracture orientations (Fig. 2a), and a 60° aperture in source-sensor paths (Fig. 3c). The second factor was determined to be about 0.8, much smaller than the minimum value of 3 used in scattering models (ACHENBACH et al., 1988), indicating that interaction between adjacent fractures is likely to occur. This interaction should contribute to the overall scattering attenuation. However, based upon the observations, we cannot determine whether this effect is frequency dependent.

The results obtained in this study suggest that, with adequate event density and distance ranges, the MLTWA approach can provide information on the relative size of scatterers present at the site. Additionally, when mapping data is limited, particularly at locations away from regions of access, MLTWA represents a reliable method for remotely characterizing the condition of the rock mass. It has been reported in the literature that observed scatterers may be related to excavation geometry (GIBOWICZ et al., 1991). As a result, the use of MLTWA may provide insight into the effect excavations have on pre-existing structures and the effectiveness of the extraction procedure.

The identification of scatterers also plays an important role in source studies, in which the scatterers may represent characteristic fault lengths that restrict the area over which rupture can proceed. Intuitively, the identified scatterers may be related to observed breakdowns in source scaling behavior, where lower levels of stress release do not correspond to similar decreases in source dimensions. In our results, the MLTWA derived characteristic fault lengths of at least 3.5 m suggest that observations of non-self-similar source scaling for source radii, representing half of the failure scale length, would occur between 1 and 2 m. URBANCIC and TRIFU (1995), using a constant Q, identified a non-similar scaling behavior for events with similar characteristic fault lengths at Strathcona mine. In their case, however, by applying the frequency dependent Q, as obtained from this study, the highest corner frequencies of small magnitude events would tend to decrease, relative to their reported values. This would result in an increase in source radii and would therefore enhance the observed non-similar behavior, further suggesting that the breakdown in scaling is influenced by the presence of characteristic faults.

Conclusions

This study has investigated the application of MLTWA to induced microseismic events ($M < -1$) for the determination of the intrinsic and scattering attenuation effects associated with an excavation at Strathcona mine, Sudbury, Canada. Within the region of monitoring, moderate stress levels exist and fracturing is considered to be relatively simple. The analyzed seismic events were recorded with an underground microseismic network encompassing the volume of interest. Insight into the behavior of scatterers was outlined in the context of rock-mass condition, excavation geometry, and source scaling relations.

The results from the attenuation analysis indicated that the intrinsic (and effective) S-wave quality factor for the Strathcona rock mass was on the order of 140 for signals with corner frequencies of around 800 Hz, whereas the scattering quality factor was about 520 at these frequencies. Our observations were found to be similar to those reported for the region based on the DSD approach and Coda-Q method. Application of MLTWA resulted in an apparent decrease in scattering for frequencies at 1600 Hz. These findings correlated well with mapped fractures and provided direct seismological evidence for minimum characteristic heterogeneity scale lengths on the order of 4 to 6 m. This suggests that MLTWA can be used to characterize the condition of the rock mass remotely when geomechanical mapping is not available or is insufficient. Additionally, if the identified scatterers are considered as characteristic fault lengths, they may play a role in any observed non-similar source scaling behavior.

Acknowledgements

Funding for this study was provided by Queen's University (A.J.F) and the Mining Research Directorate of Canada (C-I.T. and T.I.U). We graciously thank Falconbridge Ltd. for their continued support of our research efforts at Strathcona mine. We also thank Vassilios Kazakidis for valuable discussions.

REFERENCES

ACHENBACH, J. D., KITAHARA, M., MIKATA, Y., and SOTIROPOULOS, D. A. (1988), *Reflection and Transmission of Plane Waves by a Layer of Compact Inhomogeneities*, Pure and Appl. Geophys. *128*, 101–118.

AKI, K., and CHOUET, B. (1975), *Origin of Coda Waves: Source, Attenuation, and Scattering Effects*, J. Geophys. Res. *80*, 3322–3342.

BIRD, S. (1993), *Linkage of Structural Mapping, Numerical Modeling, and Microseismic Source Parameters with Application to Mine Design*, M.Sc. Thesis, Department of Mining Engineering, Queen's University, Kingston, Canada.

BRUNE, J. N. (1970), *Tectonic Stress and the Spectra of Seismic Shear Waves from Earthquakes*, J. Geophys. Res. *75*, 4997–5009. Correction in J. Geophys. Res. *76*, 5002 (1971).

CICHOWICZ, A., and GREEN, W. E. (1989), *Changes in the Early Part of the Seismic Coda due to Localized Scatterers: The Estimation of Q in a Stope Environment*, Pure and Appl. Geophys. *129*, 497–511.

CICHOWICZ, A., GREEN, W. E., VAN ZYL BRINK, A., GROBLER, P., and MOUNTFORT, P. I. *The space and time variation of microevent parameters occurring in front of an active stope*. In *Rockbursts and Seismicity in Mines* (ed. C. Fairhurst) (A. A. Balkema, Rotterdam 1990) pp. 171–175.

COATS, C. J. A., and SNAJDR, P. (1984), *Ore deposits of the North Range, Onaping-Levack Area, Sudbury*. In *The Geology and Ore Deposits of the Sudbury Structure* (eds. E. G. Pye, A. J. Naldrett and P. E. Giblin), Ontario Geological Survey, Special Volume 1, Chapter 14, 3277–346.

FEHLER, M., HOSHIBA, H., SATO, H., and OBARA, K. (1992), *Separation of Scattering and Intrinsic Attenuation for the Kanto-Tokai Region, Japan Using Measurements of S-wave Energy vs. Hypocentral Distance*, Geophys. J. Int. *108*, 787–800.

FEUSTEL, A. J. (1995), *Seismic Attenuation in Underground Mines: Measurement Techniques and Applications to Site Characterization*, Ph.D. Thesis, Department of Geological Sciences, Queen's University, Kingston, Canada.

FEUSTEL, A. J., and YOUNG, R. P. (1994), Q_β *Estimates from Spectral Ratios and Multiple Lapse Time Window Analysis: Results from an Underground Research Laboratory in Granite*, Geophys. Res. Lett. *21*, 1503–1506.

FEUSTEL, A. J., URBANCIC, T. I., and YOUNG, R. P., *Estimates of Q using the spectral decay technique for seismic events with $M < -1$*, In *Rockbursts and Seismicity in Mines* (ed. R. P. Young) (A. A. Balkema, Rotterdam 1993) pp. 337–342.

GIBOWICZ, S. J., YOUNG, R. P., TALEBI, S., and RAWLENCE, D. J. (1991), *Source Parameters of Seismic Events at the Underground Research Laboratory in Manitoba, Canada: Scaling Relations for Events with Moment Magnitude Smaller than −2*, Bull. Seismol. Soc. Am. *81*, 1157–1182.

HOSHIBA, M. (1991), *Simulation of Multiple Scattered Coda Wave Excitation Based on the Energy Conservation Law*, Phys. Earth Planet. Interiors *76*, 123–136.

HOSHIBA, M. (1993), *Separation of Scattering Attenuation and Intrinsic Absorption in Japan Using the Multiple Lapse Time Window Analysis of Full Seismogram Envelope*, J. Geophys. Res. *98*, 15809–15824.

JIN, A., MAYEDA, K., ADAMS, D., and AKI, K. (1994), *Separation of Intrinsic and Scattering Attenuation in Southern California Using TERRAscope Data*, J. Geophys. Res. *99*, 17835–17848.

MATSUMURA, S. (1981), *Three-dimensional Expression of Seismic Particle Motions by the Trajectory Ellipsoid and its Applications to the Seismic Data Observed in the Kanto District, Japan*, J. Phys. Earth. *29*, 221–239.

MAYEDA, K., SU, F., and AKI, K. (1991), *Seismic Albedo from the Total Seismic Energy Dependence on Hypocentral Distance in Southern California*, Phys. Earth Planet. Interiors *67*, 104–114.

SATO, H. (1977), *Energy Propagation Including Scattering Effect; Single Isotropic Scattering*, J. Phys. Earth. *25*, 27–41.

SCHERBAUM, F. (1990), *Combined Inversion for the Three-dimensional Q Structure and Source Parameters Using Microearthquake Spectra*, J. Geophys. Res. *95*, 12423–12438.

SNOKE, J. A. (1987), *Stable Determination of (Brune) Stress Drops*, Bull. Seismol. Soc. Am. *77*, 530–538.

SPOTTISWOODE, S. M., *Seismic attenuation in deep-level mines*. In *Rockburst and Seismicity in Mines* (ed. R. P. Young) (A. A. Balkema, Rotterdam 1993) pp. 409–414.

URBANCIC, T. I., and TRIFU, C-I. (1995), *Effects of Rupture Complexity and Stress Regime on Scaling Relations of Induced Microseismic Events*, Pure and Appl. Geophys., this issue.

WU, R. S. (1985), *Multiple Scattering and Energy Transfer of Seismic Waves: Separation of Scattering Effect from Intrinsic Attenuation, I, Theoretical Modeling*. Geophys. J. R. Astron. Soc. *82*, 57–80.

WU, R.-S., and AKI, K. (1988), *Introduction: Seismic Wave Scattering in Three-dimensionally Hetero-geneous Earth*, Pure and Appl. Geophys. *127*, 1–6.

ZENG, Y. (1991), *Compact Solution for Multiple Scattered Wave Energy in Time Domain*, Bull. Seismol. Soc. Am. *81*, 1022–1029.

ZENG, Y., SU, F., and AKI, K. (1991), *Scattering Wave Energy Propagation in a Random Isotropic Scattering Medium: 1. Theory*, J. Geophys. Res. *96*, 607–619.

(Received February 17, 1995, revised August 20, 1995, accepted August 24, 1995)

PAGEOPH, Vol. 147, No. 2 (1996)

0033–4553/96/020305–13$1.50 + 0.20/0

Variation of Certain Parameters of Regional Stress Tensor under Condition of Rockburst Hazard

Józef Dubiński[1] and Krystyna Stec[1]

Abstract —The feasibility of gaining valuable geomechanical information derived from seismological data and its specialist interpretation for utilisation in the area of assessing hazards due to mining tremors and rockbursts has become a development of signal importance in mining seismology. Undoubtedly of particular interest is a certain knowledge of the directions of the principal stresses σ_1, σ_2, σ_3 of the regional stress tensor. For their determination, use is made of a set of parameters from the mining tremors' regional focal mechanism solutions (angular parameters of nodal planes and axes of principal stresses in the tremor focus—P and T). Results of research conducted at the Szombierki and Wujek mines and analysis of calculated results for parameters of regional stress tensor show that there exist appreciable differences between values of these parameters and also a clear correlation with local extraction conditions that is of significance from the point of view of seismic hazard.

Key words: Mining seismology, rock body tremors, regional stress tensor, rockburst hazard.

1. Introduction

The mining seismology method is in practice the only one which can provide measured values of the parameters which determines the zone of formation of tremor focus, in the case of a tremor induced by mining operations, and which simultaneously permits conclusions to be drawn as to the physical processes taking place during these tremors. Carefully directed interpretation of seismological data can yield significant information, from the geomechanical point of view, which could offer a clearer insight into the nature of the mining tremors and rockbursts and hence lead to improvements in the effectiveness of prediction and also the combatting of these hazards.

The group of parameters which may be determined from this kind of interpretation includes the direction of the principal stresses σ_1, σ_2, σ_3, the so-called regional stress tensor and also the scalar magnitude R, which directly expresses the relation between these stresses. It is noteworthy that the spatial system of axes of these

---

[1] Central Mining Institute, Plac Gwarków 1, 40-166 Katowice, Poland.

stresses decides which type of tremor focal mechanism will be found. It is obviously not possible to establish absolute values of these stresses on the basis of seismological data alone.

The possibility of determining directions of the principal stresses σ_1, σ_2, σ_3 by a seismological method is of special significance, since at present neither science nor mining practice has available other comparably rapid techniques for direct assessment of stress field parameters and in particular for observation of their changes with time. The fact that these parameters correlate with local extraction conditions supplies well-based justification for considering them as precursors of changes in seismic hazard states.

2. Physical Principles of the Solutions

From the study of numerous sets of data on the seismic effects of mining origin, it is clear that the most frequently occurring focal mechanism is disruption or slip along the focal plane (GIBOWICZ, 1989). This is particularly noticeable among the higher energy tremors. The motive force for the existence of such a mechanism is a double-couple of forces and the process is of a purely shearing type. In this case the stress tensor matrix contains only components lying along its diagonal, i.e., the principal stresses σ_1, σ_2, σ_3. Hence the position of the shear plane corresponding to the focal plane will be dependent on the spatial orientation of the directions of these stresses. Contemporary seismology has at its disposal interpretative methods which make it possible to establish the position of focal plane, among which is, for example, the solution based on the signs of the first arrivals of seismic waves on the seismograms (STEC et al., 1992).

The essential feature in inversion type solutions is thus to find, based on knowledge of the parameters describing the spatial position of the focal plane, the directions of the principal stresses σ_1, σ_2, σ_3 and also the magnitude R expressing their mutual relationships.

3. Method Procedure for Determining Regional Stress Tensor

The relation between the slip vector $\check{u}$ in the fracture plane and the directions of principal stresses σ_1, σ_2, σ_3 was established by MCKENZIE (1969). The solution was derived from the following assumptions:
—the fracture process in the focus develops along a specific plane,
—the slip vector is parallel to the shear stress lying in this plane.

The stress tensor evoking this course of the destruction process may be written as follows:

$$S = \begin{bmatrix} -S_1 & 0 & 0 \\ 0 & -S_2 & 0 \\ 0 & 0 & -S_3 \end{bmatrix} \tag{3.1}$$

where $S_1 \geq S_2 \geq S_3$.

Equation (3.1) may also be written in the form:

$$S = -\begin{bmatrix} \sigma_1 & 0 & 0 \\ 0 & \sigma_2 & 0 \\ 0 & 0 & \sigma_3 \end{bmatrix} - S_3 I \tag{3.2}$$

where

$$\sigma_1 = S_1 - S_3,$$

$$\sigma_2 = S_2 - S_3,$$

$$\sigma_3 = S_3,$$

$$\sigma_1 \geq \sigma_2,$$

I—unit matrix.

After introducing the coordinates associated with the fracture plane in the focus, the form of tensor S changes to S' as in the relation

$$S' = A^{-1}SA \tag{3.3}$$

where A—unit matrix transforming any arbitrary coordinate system to a system related to the fracture plane.

Equation (3.3) may be expressed in the form

$$S'_{ij} = S_{kl} B_{jl} B_{ik} \tag{3.4}$$

where $B = A^{-1}$—matrix transforming the coordinates system of the fracture plane to a system determined by the principal axes of the stress tensor.

After introducing angles Θ, Φ, Ψ denoted as on Figure 1, the matrix B $(B = B_1 B_2 B_3)$ is the combination of successive revolutions B_1, B_2, B_3 about axes x_1, x_2, x_3 and is expressed as follows (GEPHART, 1990):

$$B = \begin{vmatrix} \cos\Psi\cos\Phi, & -\cos\Psi\sin\Phi\sin\Theta - \sin\Psi\cos\Theta, & -\cos\Psi\sin\Phi\cos\Theta + \sin\Psi\sin\Theta \\ \sin\Psi\cos\Phi, & -\sin\Psi\sin\Phi\sin\Theta + \cos\Psi\cos\Phi, & -\sin\Psi\sin\Phi\cos\Theta - \cos\Psi\sin\Theta \\ \sin\Phi, & \cos\Phi\sin\Theta, & \cos\Phi\cos\Theta \end{vmatrix}.$$

$$\tag{3.5}$$

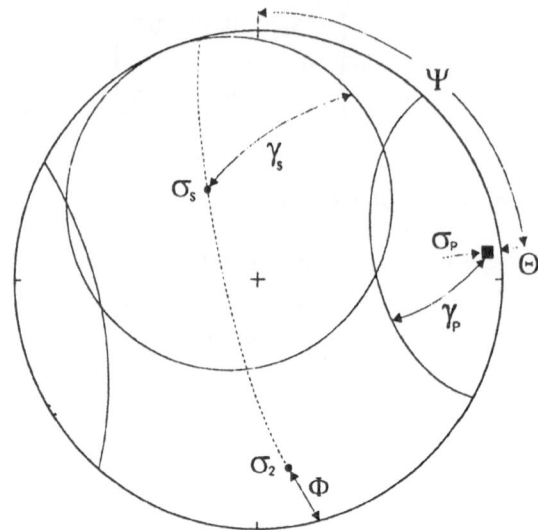

Figure 1

Stereographic projection of principal stress direction angles. Θ—plunge of axis σ_1, σ_3; Ψ—azimuth of axis σ_1, σ_3; Φ—angle of inclination of axis σ_2 with the plane containing axes σ_1, σ_3; γ_s, γ_p—range of variation of directions of determined stresses about the initial estimates σ_p, σ_s.

Ultimately the following form is obtained:

$$S'_{13} = -\sigma_1 B_{11} B_{13} - \sigma_2 B_{12} B_{23}$$

$$S'_{13} = -[\sigma_1 \cos \Psi \sin \Phi - \sigma_2 \sin \Theta (\cos \Psi \sin \Phi \sin \Theta + \sin \Psi \cos \Theta)] \cos \Phi. \quad (3.6)$$

Equation (3.6) permits obtaining of a solution for the principal directions of a triaxial stress tensor when the following conditions are satisfied

$$\frac{\sigma_1}{\sigma_2} = \sin^2 \Theta + \frac{tg\ \Psi}{\sin \Phi} \sin \Theta \cos \Theta. \quad (3.7)$$

The shear stress S'_{12} then has the value

$$S'_{12} = -\sigma_1 \sin \Psi \cos \Psi \cos^2 \Phi - \sigma_2 (\cos \Psi \sin \Phi \sin \Theta + \sin \Psi \cos \Theta)$$

$$\cdot (\sin \Psi \sin \Phi \sin \Theta - \cos \Psi \cos \Theta). \quad (3.8)$$

The calculation procedure described here is of the nature of an algorithm for determining the directions of the stress tensor causing slip in the positive direction of axis x_2 of the fracture plane of the focus. Based on these calculations, a computer programme FMSI (GEPHART, 1990) was developed for the calculation of selected parameters of the regional stress field. Among them are the azimuth and plunge angles which determine the directions of axes of principal stresses σ_1, σ_2, σ_3 and also the scalar magnitude R (ETCHECOPAR et al., 1981), where

$$\frac{\sigma_2 - \sigma_1}{\sigma_3 - \sigma_1} = -\frac{\beta_{13}\beta_{23}}{\beta_{12}\beta_{22}} = R \tag{3.9}$$

where β_{ij}—cosines of angles between the coordinates of the system x_1, x_2, x_3 associated with the directions of principal stresses and the coordinates of the system x'_1, x'_2, x'_3 associated with the fracture plane in the focus.

When making the assumption $\sigma_1 \geq \sigma_2 \geq \sigma_3$ the parameter R takes the value $0 \leq R \leq 1$. The data base for calculation of the magnitudes given above is formed by the parameters of tremors focal mechanism solutions. Depending on the set of parameters of the focal mechanism, six alternative data files may be derived:

1. Angular parameters of two nodal planes (azimuth, dip) and slip vector direction taking into account weighted values,

2. Angular parameters of axes of principal stresses P and T (azimuth, plunge),

3. Angular parameters of fracture plane (azimuth, dip) and also azimuth and plunge of the slip vector direction,

4. Angular parameters of fracture plane (azimuth, dip) and also the sign of slip vector direction ("+" or "−"),

5. Angular parameters of fracture plane (azimuth, dip) and also the value of slip vector index (from 1 to 4),

6. Angular parameters of fracture plane (azimuth, dip) and also the value of slip angle.

The weighted values depend on the energy of seismic events and the accuracy of the solution of their mechanisms. The sign of the slip vector implies the direction of dip slip ("+" is taken for normal model of focus; "−" for inverted model of focus). Designation of the slip direction is as follows: 1—for events with normal dip slip, 2—for events with inverted dip slip, 3—for events with dextral strike slip, 4—for events with sinistral strike slip.

Results of calculations of regional stress tensor are presented in the tables, including the error expressing the mean misfit for values R and Φ. Table 1 is an example of this. The form of presentation of results is very important for a realistic evaluation of their reliability and the feasibility of determining allowable models. As the final solution is taken, the model with the least sum of errors, for which, after reading off R and Φ, are determined the angular parameters (azimuth, plunge) of directions of axes of the principal stresses σ_1, σ_2, σ_3.

4. Calculation Examples

Calculations were performed making use of the FMSI programme which made it possible, based on data formed by angular parameters of the tremors' focal mechanism solutions, to determine the directions (azimuth, plunge) of principal stresses σ_1, σ_2, σ_3. Research was carried out in the Szombierki mine, longwalls 21E

Table 1

Sum of misfit errors for stress parameters σ_1, σ_2, σ_3

	R								
Φ	5.9	14.9	23.9	41.9	50.9	59.9	68.9	77.9	86.9
0.00	82.58*	62.049	54.270	35.358	34.623	40.966	48.210	54.735	60.656
0.10	20.001	27.590	27.511	21.322	20.206	24.859	30.249	34.317	37.829
0.20	25.681	30.072	29.229	19.856	25.029	27.329	30.782	34.915	39.058
0.30	32.005	24.392	25.184	17.928	25.946	29.061	31.160	35.293	39.935
0.40	30.744	21.948	21.711	17.370	26.632	29.498	31.142	35.351	39.521
0.50	29.629	21.164	20.685	18.003	26.607	29.247	30.532	34.952	37.900
0.60	28.607	20.837	20.228	18.570	26.499	27.786	28.591	32.490	34.734
0.70	27.692	20.732	20.094	19.145	26.489	27.040	26.233	27.912	29.042
0.80	26.882	20.751	20.185	19.582	26.678	26.476	25.439	24.259	42.396
0.90	26.181	20.866	20.490	20.133	26.300	25.739	25.010	24.401	32.807
1.00	57.480	57.480	57.480	57.480	57.480	57.480	57.480	57.480	57.480

R—Etchecopar's parameter, Φ—inclination angle σ_2, *Sum of errors is expressed in degrees.

and 22E, and in the Wujek mine, longwalls 11b and 12b. The relevant geological and mining situation is shown in Figures 2 and 3.

In order to display regularities in the space–time variations of the chosen regional stress tensor parameters, the set of input data was systematised according to the criteria of position of the tremor focus relative to the extraction front. Calculations were made for monthly extraction periods, distinguishing groups of events situated in advance of, and behind the longwall front

As a result of applying this procedure, for each data set a series of results tables was obtained (Table 2 is an example) containing:

—angular parameters of directions of principal stresses σ_1, σ_2, σ_3 (azimuth $\varphi_{1,2,3}$, plunge $\delta_{1,2,3}$),
—coefficient R,
—error in angles' determination.

5. Analysis of Results

An analysis of calculated results obtained for angular parameters of principal stresses of the regional stress tensor and for coefficient R determined from the set of parameters of the mining tremors focal mechanisms' solutions, shows a clear differentiation in the test regions and also shows a correlation with changes in the stress-strain state in the rock mass induced by mining operations.

Results from the Szombierki mine for the region of longwalls 21E and 22E demonstrate that the distribution of directions of stresses σ_1, σ_2, σ_3 and parameter R change together with the level of seismic activity described by parameter b of the

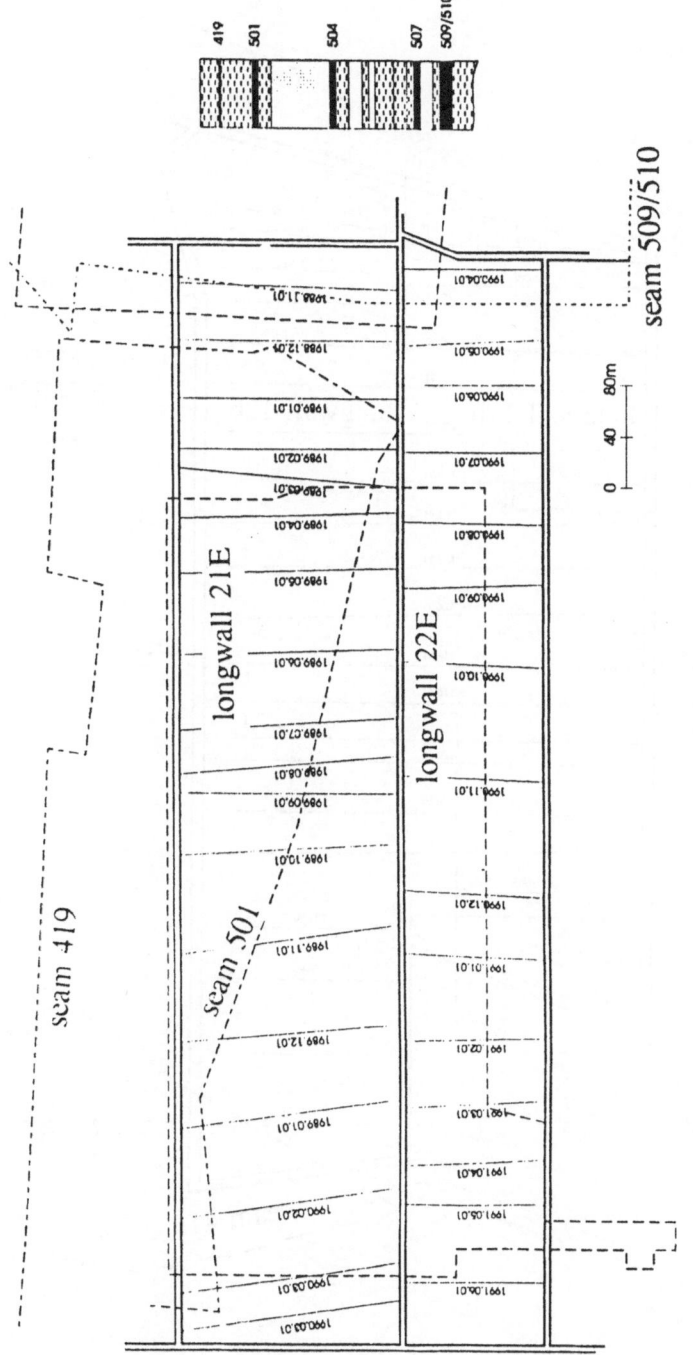

Figure 2

Geological and mining situation in the region of longwalls 21E and 22E at the Szombierki mine.

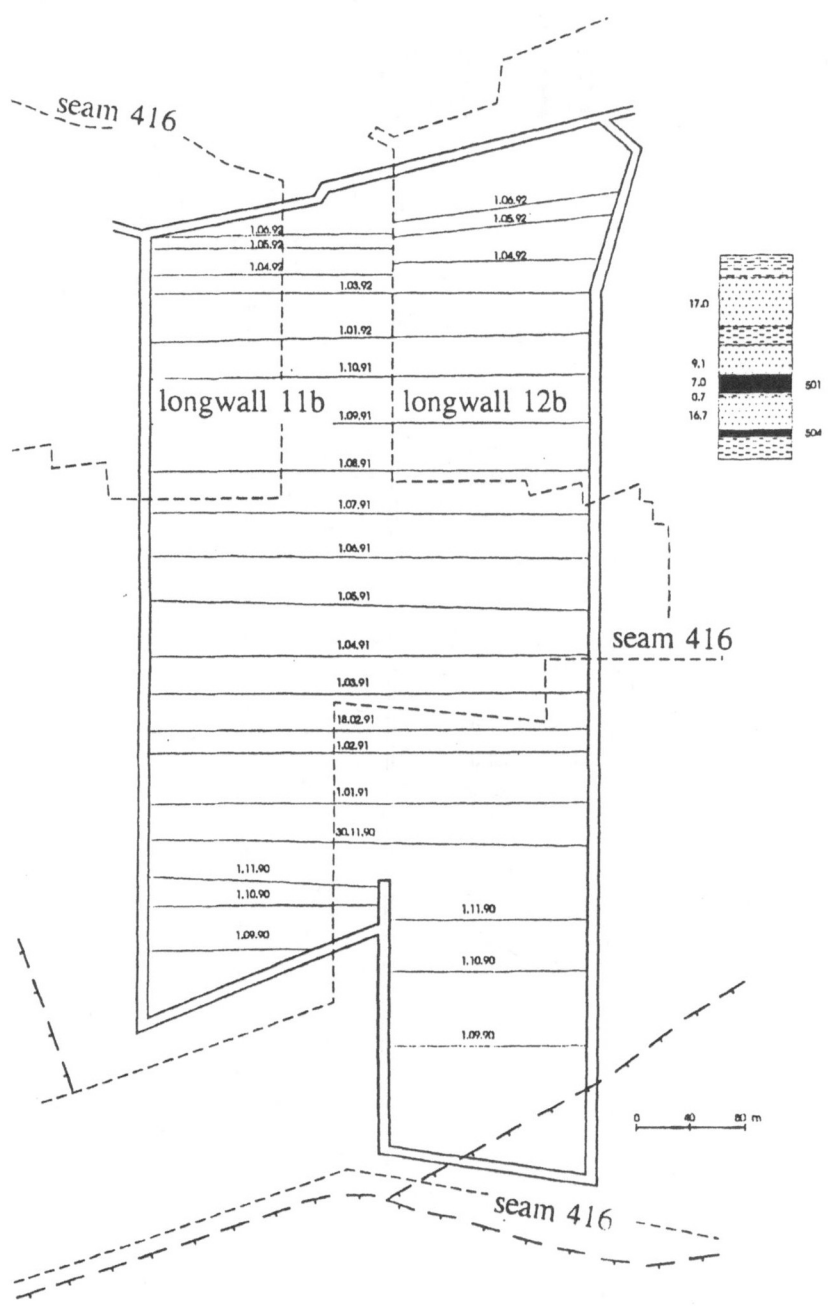

Figure 3
Geological and mining situation in the region of longwalls 11b and 12b at the Wujek mine.

Table 2

Results of determination of axes directions of regional stress tensor

Szombierki mine, longwall 21E
Foci of mining tremors which occurred ahead of the face

Time	σ_1		σ_2		σ_3		R	Error [°]
	δ_1	φ_1	δ_2	φ_2	δ_3	φ_3		
10	32	287	58	98	4	149	0.2	6
11	19	272	68	62	10	178	0.4	5
12	17	268	73	85	3	176	0.4	7
1	67	308	11	92	12	186	0.7	5
2	51	276	17	29	34	131	0.6	6
3	26	200	28	95	50	325	0.5	7
4	25	183	15	91	65	249	0.8	4
5	27	201	23	94	51	328	0.7	9
6	44	201	27	82	35	332	0.7	8
7	69	148	7	39	19	307	0.4	6
8	19	269	67	60	11	180	0.4	5
9	18	266	72	86	1	178	0.3	4
10	6	51	21	144	68	396	0.6	5
11	7	25	53	123	38	289	0.5	6
12	40	193	26	78	39	326	0.4	7
1	17	222	54	107	31	322	0.2	7
2	27	121	63	283	3	19	0.1	4
3	28	124	42	242	30	11	0.3	6
4	18	115	69	315	8	203	0.4	7

Gutenberg-Richter distribution and by parameter ERR expressing the quantity of energy released per unit area of exposed roof (Fig. 4).

In the initial period of extraction of longwall 21E (XI/89) low values of parameter R (0.9) were observed, giving evidence that at that time a compression process predominated, whereas since the following month the value of R has increased, indicating the appearance of tensile processes. These processes are also reflected in the change in parameters of the focal mechanism. At this time the longwall extraction front lay underneath the unmined part of seam 504 and the unmined part left in seam 501. At this time the seismic activity was high.

Next, the R parameter found over successive months of extraction of longwall 21E shows values greater than 0.5, which indicates the presence of increased tensile stress in the rock mass. For the majority of tremors throughout the whole period, the angular parameters of P and T axes are similar except for the month of June 1989, when a very clear change in direction of stress T was observed. This change was confirmed by the occurrence of a very severe tremor of energy $E = 3 \cdot 10^8$ J, which caused the formation of a new stress distribution, lasting through the following months to the end of extraction of longwall 21E. Throughout virtually

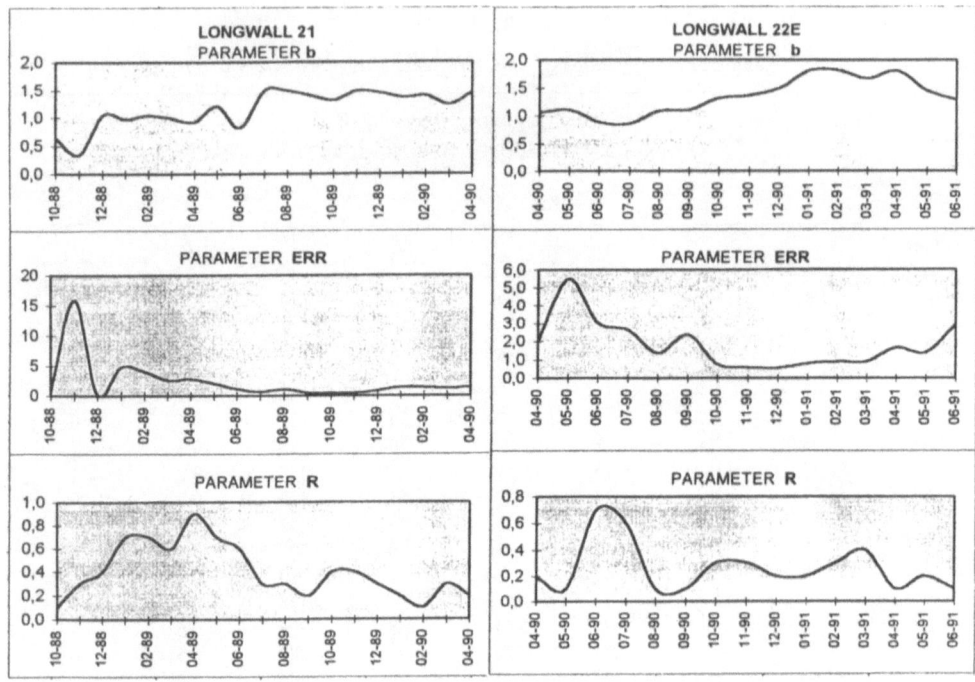

Figure 4
Distribution of parameters b, ERR, and R during the mining of longwalls 21E and 22E.

the entire period the parameter R exhibited values lower than 0.5 (compressive stresses predominating) with the exception of October 1989 ($R = 0.6$). An increase in value of parameter R correlated here with the passing of the longwall extraction front under the edges of overlying seams.

The analysis for longwall 21E presented here may be seen to indicate a clear relationship between the conditions of extraction in this longwall and changes in the distribution of stress field, the occurrence of focal mechanisms of various types and also the level of seismic activity. The characteristic features of this relationship are as follows:

1. During the period when the longwall extraction front was passing through the area under mined out seams, compression stresses predominated in the rock body. The value of parameter R was less than 0.5. The principal focal mechanism was the dip-slip normal mechanism with focal motion approximately vertical.

2. When the extraction front was passing under unmined parts of the rock mass and also in places where the influence of edges was shown, tensile stresses ($R > 0.5$) were chiefly found. The most frequent type of focal mechanism was the slip with horizontal motion in the focus.

3. The occurrence of increased seismic activity was observed, manifested by a greater quantity of seismic events during the time when parameter R was achieving

values greater than 0.5, i.e., during the period when tensile stresses dominate. In the geomechanical interpretation of the presented results it should be noted that induced seismic events resulting from extraction of the longwall and local stress-strain states (e.g., extraction edges) are associated with a certain displacement, due to the stress distribution ahead of the advancing longwall front.

In the case of longwall 22E, calculated results for regional stress tensor also show several characteristic features. In longwall 22E, throughout virtually the entire extraction time, compression stresses predominated ($R < 0.5$). The first change in stress distribution occurred in June and July of 1990. The parameter R then reached a value of the order of 0.6–0.7. This change may be associated with the occurrence of extraction edges in seams 501 and 504.

The analysis conducted indicates that the occurrence of values of parameter R close to 1 is linked with an increased level of seismic activity.

An analysis of parameters of regional stress tensor was also performed for the set of temors' mechanism solutions in the vicinity of longwalls 11b and 12b at the Wujek mine (Fig. 5). It was found that during the period when the extraction front of longwall 11b was passing under the mined out part of seam 416 (June–July, 1990) low R values of the order of 0.2 occurred, giving evidence of distinct

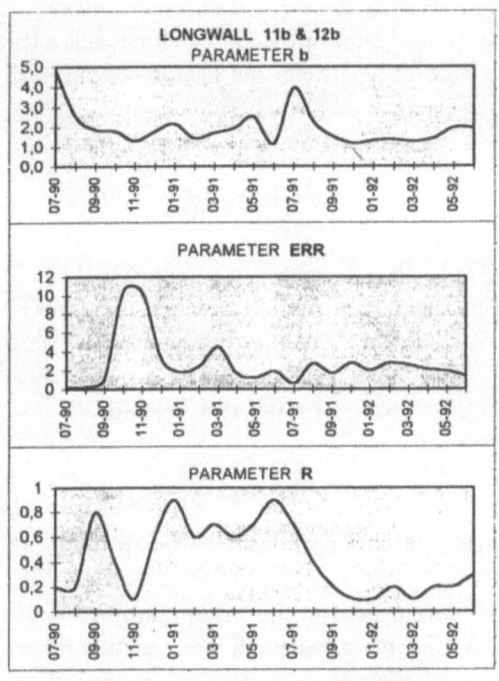

Figure 5
Distribution of parameters b, ERR, and R during the mining of longwalls 11b and 12b at the Wujek mine.

predomination of compressive stresses. Foci of tremors which occurred in that period exhibited mainly slip-inverted mechanism. In September 1990 a very marked rise in parameter R to a value of 0.8 was noted. This high value of R indicated a strong domination of tensile stresses most probably caused by an extraction edge in seam 416 parallel to the longwall front.

During the next period of extraction of longwalls 11b and 12b (October, 1990), R had a value of about 0.4, dropping further to 0.1 in November, 1990. The occurrence of high tensile stresses in that period was found when longwall 11b was being worked under the unmined part of seam 416 and longwall 12b under the mined out part.

During the following extraction period (up to the middle of July, 1991) the extraction fronts of longwalls 11b and 12b were passing under the unmined section of seam 416. This situation is reflected in the stress distribution and the focal mechanism. At that time, parameter R took a value greater than 0.5, indicating the domination of tensile stresses. The principal type of focal mechanism occurring at that time was the mechanism of slip with a horizontal motion of the ground in the focus, i.e., the strike-slip mechanism. From mid-July, 1991 both longwalls were worked under the safety pillar protecting the main cross-cuts left in seam 416. The parameter R took low values, that is compression predominated. Tremors occurring in this period mostly exhibited the normal slip mechanism.

In general, it may be said that in the case of longwalls 11b and 12b a very clear correlation could be observed between the heightened level of seismic activity and the domination of tensile stresses ($R > 0.5$).

5. Conclusions

1. An analysis of selected parameters of the regional stress tensor, calculated on the basis of a set of parameters of mining tremors' focal mechanisms solutions, has demonstrated the existence of a distinctly differentiated stress field within the tested regions.

2. The parameter R expressed by the relationship:

$$R = \frac{\sigma_2 - \sigma_1}{\sigma_3 - \sigma_1}$$

carries valuable information and correlates well with the local extraction condition and seismic hazard.

3. The procedure developed for the interpretation of seismological data may serve to widen the range of parameters and enable fuller use of information contained in the tremors' seismograms. This procedure offers great possibilities of practical application and also the feasibility of achieving improvements in the effectiveness of rockburst hazard seismological assessment.

REFERENCES

AKI, K., and RICHARDS, P. G., *Quantitative Seismology – Theory and Methods*, vols. 1, 2 (W. H. Freeman and Co., San Francisco 1980).

BRILLINGER, D., UDIAS, A., and BOLT, B. A. (1980), *A Probability Model for Regional Focal Mechanism Solutions*, Bull. Seismol. Soc. Am. *70*, 149–170.

ETCHECOPAR, A., VASSEUR, G., and DAIGNIERS, M. (1981), *An Inverse Problem in Microtectonics for the Determination of Stress Tensors from Fault Striation Analysis*, J. Struct. Geol. *3*, 51–65.

GEPHART, J. W., and FORSYTH, D. W. (1984), *An Improved Method for Determining the Regional Stress Tensor Using Earthquake Focal Mechanism Date: Application to the San Fernando Earthquake Sequence*, J. Geoph. Res. *89*, 9305–0320.

GEPHART, J. W. (1990), *FMSI: A Fortran Program for Inverting Fault-slickenside and Earthquake Focal Mechanism Data to Obtain the Regional Stress Tensor*, Computers and Geosciences *16* (7), 953–989.

GIBOWICZ, S. J. (1989), *Mechanizm ognisk wstrząsów górniczych*, Publ. Inst. Geoph. Pol. Ac. Sc. M–13 (221), PWN, Waszawa-Lódź.

KASAHARA, K., *Earthquake Mechanism* (Cambridge Univ. Press., London 1981).

McKENZY, D. P. (1969), *The Relation between Fault Plane Solution and the Directions of the Principal Stresses*, Bull. Seismol. Soc. Am. *59*, 591–601.

STEC, K., DUBIŃSKI, J., and NOWAK, J. (1992), *Correlation between the Parameters of Mining Tremors Focal Mechanism and the Seismic Hazard State Based on an Example of the Wujek Coal Mine*, Acta Montana *2* (88), 145–160.

WONG, I. C. (1993), *Tectonic stresses in mine seismicity: Are they significant?* In *Rockburst and Seismicity in Mines 93* (ed. R P. Young) (Kingston, Ontario 1993) pp. 273–278.

(Received January 16, 1995, revised October 12, 1995, accepted October 16, 1995)

PAGEOPH, Vol. 147, No. 2 (1996)

0033–4553/96/020319–25$1.50 + 0.20/0

Effects of Rupture Complexity and Stress Regime on Scaling Relations of Induced Microseismic Events

THEODORE I. URBANCIC[1] and CEZAR-IOAN TRIFU[1]

Abstract—Source parameter scaling relations are examined for microseismic events ($-2.4 \leq M \leq -0.3$) occurring within highly and moderately stressed and fractured rock masses at Strathcona mine, Sudbury, Canada. Insight into scaling is provided by waveform complexities, calculated rupture velocities, and maximum shear stresses based on *in situ* and numerical modelling data. The importance of normal stress on the failure process is also considered. Our results show that a strong dependence exists between stress release and seismic moment. An observed positive scaling in excess stress release ($\Delta\sigma/2 - \sigma_a$) is consistent with the concept of overshoot. Rupture velocities ranging from 0.2 to 0.5β and waveform complexities less than 1.5 suggested that overshoot was related to healing behind a slowly advancing rupture front. Scaling in seismic efficiency paralleled that in apparent stress, implying that seismic stress release estimates are quasi-independent of the maximum shear stress. High levels of normal stress further supported the importance of high resisting stress in the observed overshoot behaviour and its role in the failure process.

Key words: Microseismicity, source parameter scaling, seismic efficiency, principal stresses.

Introduction

Traditionally, scaling relations consider how stress release changes when the source dimensions increase with increasing magnitude or seismic moment. Numerous studies have shown that globally, and over a wide magnitude range, the stress release does not depend on the magnitude (self-similar behaviour). For example, static stress drops tend to vary from 0.1 to 10 MPa for seismic moments ranging from 10^4 to 10^{22} N.m (e.g., HANKS, 1977). However, on local or regional scales, and over limited magnitude ranges, non-similar scaling behaviour in stress release has been observed. This has been intepreted to be either related to the source, involving the presence of characteristic fault dimensions (AKI, 1984), or to any process that limits high frequencies, such as anelastic attenuation and site recording effects (FLETCHER *et al.*, 1986).

Scaling behaviour has implications on how smaller magnitude events may be related to the occurrence of larger and more damaging earthquakes. In addition,

---

[1] Engineering Seismology Group Canada Inc., Kingston, Canada, K7L 2Z4.

scaling also provides information on the processes associated with earthquake generation. In order to investigate the effects responsible for different scaling behaviours, it is highly desirable to be closer to the source itself. Such opportunities are provided by underground mining environments where seismic networks can be installed to encompass the sources of seismicity and additional information on the rock mass, for example fracture characteristics and stress conditions (orientation and values), can be obtained through direct measurements. As a result, this environment may allow better insight into possible relationships between parameters calculated using seismic waveforms and rock mass conditions.

For mining-induced studies, both self-similar and non-similar scaling behaviours have been reported. MCGARR (1984) considered that both depth and tectonic regime (compressional or extensional) influence scaling. Conversely, CICHOWICZ et al. (1990) proposed that inappropriate attenuation corrections are responsible for non-similar scaling. Further, it was suggested by GIBOWICZ et al. (1991) that non-similar scaling in mines reflects variations in fracturing and source size related to the stress release around openings. URBANCIC et al. (1992) considered that the interaction of stresses with pre-existing fractures, fracture complexity, and depth of events are mainly responsible for observed scaling behaviour. They also speculated that self-similar relationships were characteristic for events with similar depths or for weakly structured rock masses with reduced clamping stresses, whereas non-similar behaviour was characteristic for events with variable depth or for heavily fractured zones under stress confinement.

In this study, we expand on the investigation into scaling behaviour by examining the role of source complexity, rupture velocity, and deviatoric stress. Our analysis considers microseismicity ($-2.4 \leq M \leq -0.3$) occurring within both highly and moderately stressed and fractured rock masses at Strathcona mine, Sudbury, Canada. Scaling relations are derived based on the spectral analysis of seismic signals, whereas source complexity and rupture velocity are evaluated from seismic waveform analysis. Underground structural mapping and *in situ* stress measurements, in conjunction with three-dimensional numerical modelling of stresses are used to determine the maximum shear stress. We then examine the seismic efficiency as a function of seismic moment. The importance of normal stress on the failure process and consequently on scaling behaviour is also outlined.

Site Conditions

Strathcona mine is a nickel-copper mine located in the Sudbury Basin (Ontario). The main orebody lies in a sulphide rich unit, dipping at 45° to 80° to the south with a north-northeast strike, between a mafic norite unit above and a granitic-gneiss unit below (NALDRETT and KULLERUD, 1967). The two excavation sites we investigate for associated seismicity are located in geologically diverse domains at

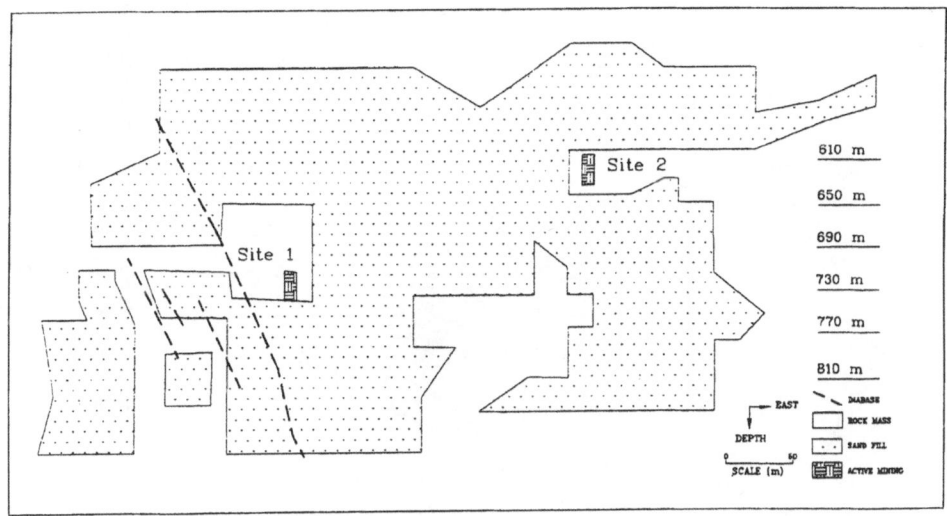

Figure 1
Schematic longitudinal view of Strathcona mine, along the strike of the orebody, showing the location
of the two study sites.

depths of about 600 to 750 m (Fig. 1). Site 1 consists of a complex orthorhombic
pattern of fractures as determined from structural mapping (Fig. 2a). Conversely,
Site 2 is dominated by a single east-west trending structural set (Fig. 2b). For both
sites, all fractures with lengths greater than 1.5 m within the volume surrounding
the excavations were mapped (143 and 292, respectively). Fracture spacing for the
sites varied from 0.3 to 2.0 m, with the largest percentage at about 1 m. Most of the
surfaces for these fractures are rough and undulating, and are coated with a thin
layer of chlorite (BIRD, 1993). Additional fracture characteristics for Site 2, such as
compressive strength, roughness coefficient, and residual friction angle were measured and input into the numerical modelling for the calculation of stress distribution within the monitoring volume.

The extraction of ore at the sites was carried out in successive stages over six
month periods in 1987 and 1991–1992, respectively, and final extraction volumes
were on the order of 7000 m³. Site 1 was located towards the centre of the mine,
where stress levels were elevated as a result of extensive past mining activity. *In situ*
stress measurements, based on overcoring tests, indicated that the magnitudes of
the principal stresses were 60, 56, and 32 MPa. The maximum and intermediate
principal stresses were oriented horizontally, to the northwest and northeast,
respectively. Site 2 was located at the northeast edge of the mine, in an area more
directly influenced by the regional stress field. The average magnitudes of the
far-field principal stresses, in agreement with underground measurements, were 27,
23, and 16 MPa (COCHRANE, 1991). As above, the maximum and intermediate

Theodore I. Urbancic and Cezar-Ioan Trifu PAGEOPH,

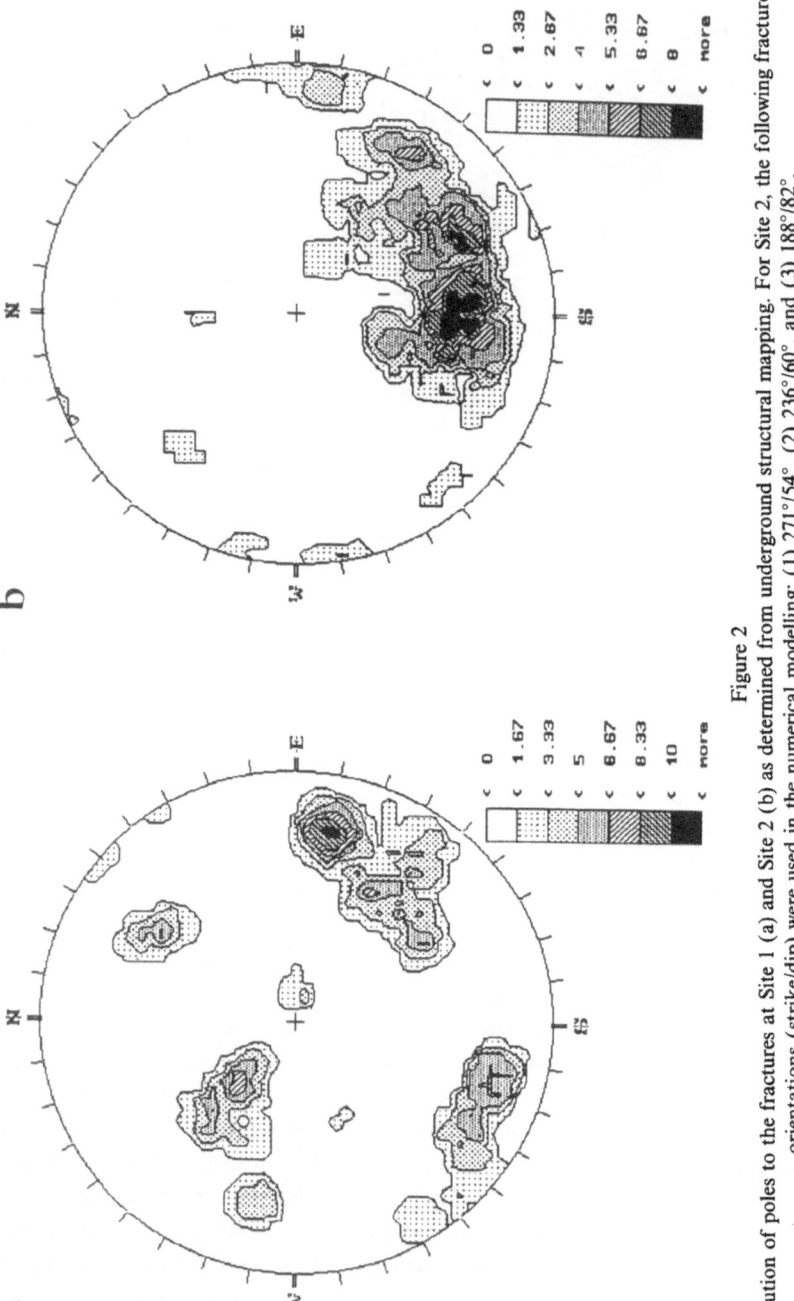

Figure 2

Distribution of poles to the fractures at Site 1 (a) and Site 2 (b) as determined from underground structural mapping. For Site 2, the following fracture set orientations (strike/dip) were used in the numerical modelling: (1) 271°/54°, (2) 236°/60°, and (3) 188°/82°.

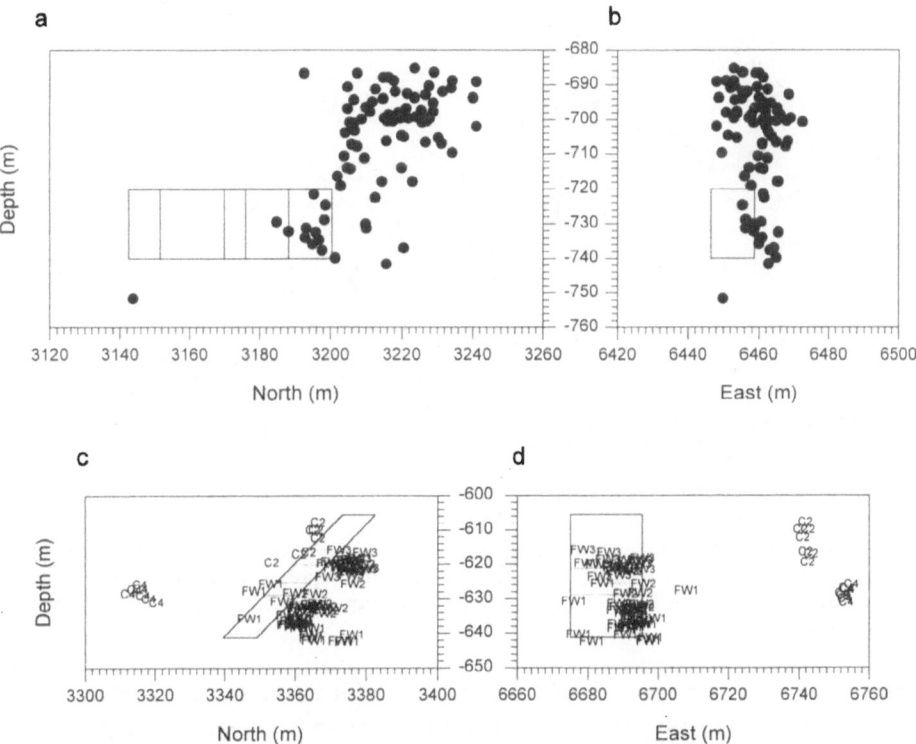

Figure 3
Scaled representations of event locations and excavations at Site 1 (a, b) and Site 2 (c, d). For Site 2, the symbols represent different event clusters that will be referred to in the analysis. Extraction stages are also indicated, advancing from south to north at Site 1, and toward shallower depths at Site 2.

principal stress orientations are horizontal, however, they trend east-west and north-south, respectively.

Data Analysis

The data analyzed for Site 1 (Fig. 3a) was collected between August and December, 1987 corresponding to <25%, 50–60%, and 100% extraction. The Site 2 analysis (Fig. 3b) includes the examination of seismicity following progressive stages of excavation, between October 1991 and January 1992, including a period of non-mining (shutdown) during December, 1991 (25% of analyzed events at this site). Seismic monitoring of the two sites was achieved with 32 and 64 channel full waveform microseismic systems, respectively. Each network employed 5 triaxial accelerometers, single and/or dual gain, and additional uniaxial accelerometers (17 and 33, respectively). The sensors were optimally distributed in three dimensions

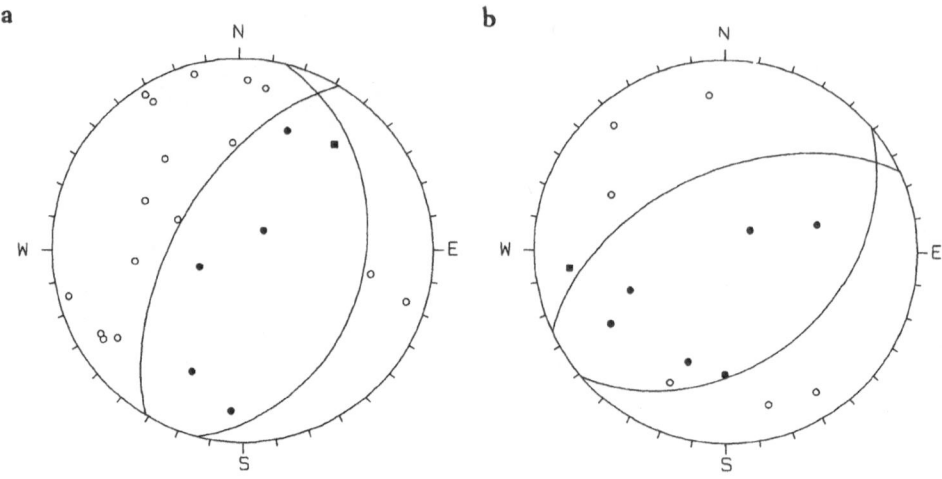

Figure 4
Examples of fault-plane solutions, as plotted on lower hemisphere equal area stereonets, for events recorded at Site 2. Compressional first P-wave motions are represented by filled circles whereas dilatational first motions by open circles; uncertain polarity readings are indicated by squares.

with an inter-sensor spacing of about 50 to 100 m for Site 1 and 30 to 60 m for Site 2. Calibration surveys were performed to determine the system response and analog-to-digital conversion was carried out with 12-bit resolution at a sampling rate of 20 kHz. Only signals with signal-to-noise ratios greater than 4 were used for locating events; this allowed for accurate arrival-time picks to better than 0.1 to 0.15 msec. Events were located based on a minimum of 10 P-wave arrivals; using a combined Simplex/Geiger algorithm and incorporating average P-wave velocities obtained by active velocity surveys ($\alpha = 6197$ and 5970 m/s, respectively). Location errors (in a vectorial sense) were typically 5 to 6 m for Site 1 and 2 to 3 m for Site 2. Only events with unobscured ray paths to at least 3 triaxial sensors and with fault-plane solutions that fit the double-couple model were considered for detailed analysis (85 and 68; Fig. 4). Additionally, by assuming that a reasonably uniform stress field existed in the regions defined by seismicity, the stress inversion technique developed by GEPHART and FORSYTH (1984) was used to identify the fault plane in focal mechanism solutions.

Source parameters were estimated using the triaxial sensors which had flat frequency responses (within ± 3 dB) between 50 Hz and 5 kHz. The effective frequency bandwidth for Site 2 ranged from 100 Hz to 3.5 kHz due to the characteristics of the employed pre-amplifiers. The analysis was carried out on rotated P- and S-wave signals. In the frequency domain, anelastic attenuation and scattering effects were taken into account by multiplying the spectra with $\exp(\pi f t^*)$, where f is frequency, t^* is R/vQ_v, R is the source-sensor separation, v is either the

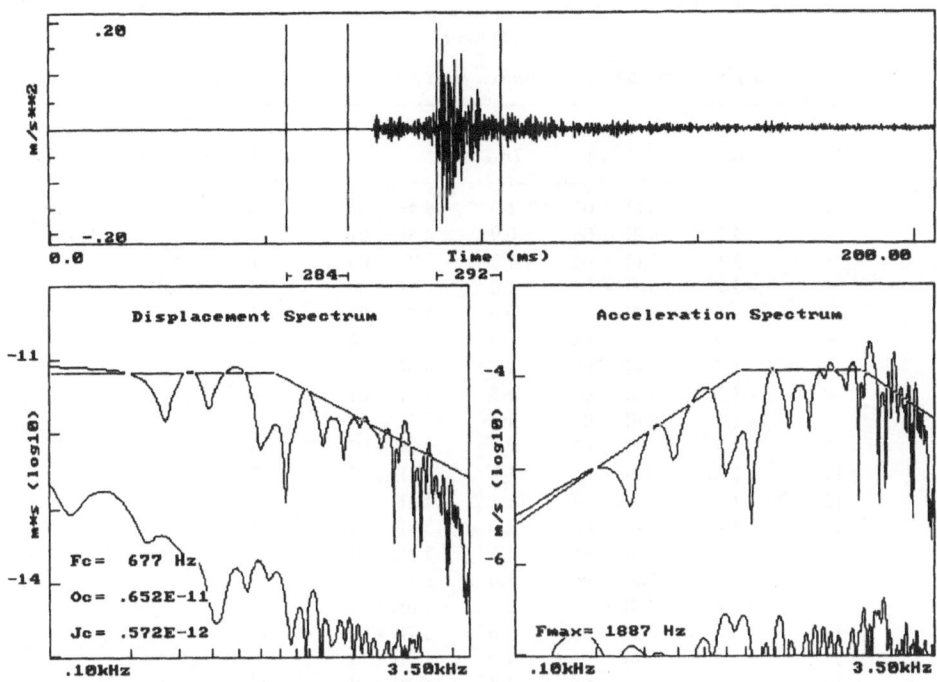

Figure 5

Example of a rotated S-wave signal (Site 2), showing the signal and noise windows, and the corresponding displacement and acceleration spectra, where F_c is the corner frequency, O_c is the spectral level and J_c is the energy flux.

P- or S-wave ($\beta = 3700$ m/s) velocity, and Q_v is the average quality factor along the ray path, ranging for both P and S waves from 100 to 150 (FEUSTEL et al., 1993). By applying the above Q_v values, the spectral decay was generally well described by a -2 slope (Fig. 5). For Site 2, time domain studies were also carried out using rotated P- and S- wave acceleration and integrated velocity signals. Measurements included peak and rms accelerations, S-wave first pulse durations, rise times, and overall signal durations. Peak accelerations were corrected for attenuation and scattering by multiplying with $\exp(\pi f_a t^*)$, where f_a is the calculated corner frequency that would correspond to the most energetic asperity. Corrections for rms accelerations, initial pulse durations and rise times and overall pulse durations were made by subtracting t^*, $t^*/2$ and $3t^*/2$, respectively.

Determination of Source Parameters

Several source parameters were estimated from spectral analysis (Tables 1 and 2). The seismic moment M_0 was evaluated by

$$M_0 = \frac{4\pi\rho v^3 R |\Omega_{0v}|}{F_v} \tag{1}$$

Table 1

Source parameters as determined from spectral analysis for Site 1

No.	M	M_0 (N.m)	r_0 (m)	E_0 (J)	E_s/E_p	η
1	−2.1	6.6E + 05	1.0	6.4E − 01	3	−
2	−2.2	4.9E + 05	0.9	4.4E − 01	2	−
3	−2.1	7.6E + 05	1.3	1.4E + 00	7	−
4	−2.2	5.5E + 05	1.0	1.7E + 00	20	8.0E − 04
5	−2.1	8.3E + 05	0.9	1.8E + 00	2	−
6	−1.9	1.2E + 06	0.9	4.1E + 00	5	−
7	−1.5	4.8E + 06	1.2	8.6E + 01	29	5.1E − 03
8	−1.9	1.6E + 06	1.0	1.1E + 01	6	−
9	−1.1	2.6E + 07	1.1	1.5E + 03	8	−
10	−1.2	1.5E + 07	2.0	1.7E + 02	3	−
11	−2.0	9.1E + 05	1.0	1.3E + 00	2	−
12	−1.3	1.1E + 07	1.1	1.4E + 02	3	−
13	−1.4	8.4E + 06	2.0	3.7E + 01	1	−
14	−1.7	3.2E + 06	1.6	1.8E + 01	9	1.6E − 03
15	−1.5	5.3E + 06	1.0	3.2E + 01	2	−
16	−1.6	4.4E + 06	1.1	7.0E + 01	3	−
17	−1.7	3.2E + 06	1.0	3.7E + 01	10	3.3E − 03
18	−1.6	4.4E + 06	1.3	1.3E + 01	3	−
19	−1.5	5.1E + 06	1.4	4.7E + 01	13	2.5E − 03
20	−0.9	4.0E + 07	1.3	1.5E + 03	2	−
21	−1.2	1.8E + 07	2.0	3.2E + 02	14	5.0E − 03
22	−1.3	1.3E + 07	1.2	1.1E + 02	1	−
23	−1.7	3.3E + 06	1.9	9.6E + 00	4	−
24	−1.6	3.9E + 06	1.5	7.1E + 00	1	−
25	−1.6	4.5E + 06	1.2	1.5E + 01	2	−
26	−1.4	9.2E + 06	1.4	6.1E + 01	1	−
27	−1.6	4.4E + 06	1.9	1.7E + 01	8	−
28	−1.2	1.6E + 07	1.7	1.2E + 02	6	−
29	−0.7	1.0E + 08	2.2	1.8E + 03	2	−
30	−1.0	3.7E + 07	1.8	1.8E + 03	101	1.3E − 02
31	−1.0	3.6E + 07	1.3	1.6E + 03	10	1.2E − 02
32	−0.9	5.2E + 07	2.1	8.8E + 03	38	4.8E − 02
33	−1.3	1.2E + 07	1.5	1.1E + 02	3	−
34	−0.6	1.2E + 08	2.3	1.2E + 03	1	−
35	−1.0	3.3E + 07	1.5	2.1E + 03	47	1.8E − 02
37	−0.7	9.5E + 07	1.6	9.9E + 03	14	2.9E − 02
37	−1.3	1.0E + 07	1.4	4.7E + 01	1	−
38	−0.7	7.8E + 07	1.6	1.7E + 03	1	−
39	−0.8	5.5E + 07	1.4	3.3E + 03	6	−
40	−0.6	1.4E + 08	1.5	6.9E + 03	2	−
41	−0.6	1.2E + 08	1.5	3.8E + 03	1	−
42	−1.0	3.6E + 07	1.3	1.1E + 03	4	−
43	−1.0	2.8E + 07	2.1	2.3E + 02	11	2.2E − 03
44	−1.1	2.2E + 07	1.2	1.1E + 03	17	1.4E − 02
45	−0.7	8.6E + 07	1.5	6.4E + 03	2	−
46	−1.4	8.2E + 06	1.7	2.0E + 01	2	−
47	−1.5	6.3E + 06	1.6	5.4E + 01	25	2.4E − 03

Table 1 (*Contd*)

No.	M	M_0 (N.m)	r_0 (m)	E_0 (J)	E_s/E_p	η
48	−0.4	2.3E + 08	1.3	2.8E + 04	1	−
49	−1.1	2.4E + 07	1.9	1.9E + 02	8	−
50	−0.9	3.9E + 07	1.4	1.3E + 03	5	−
51	−1.0	2.8E + 07	1.8	1.7E + 02	2	−
52	−1.0	3.3E + 07	1.7	5.4E + 02	7	−
53	−0.5	1.6E + 08	2.3	1.3E + 04	51	2.2E − 02
54	−1.2	1.7E + 07	1.5	2.5E + 02	6	−
55	−1.1	2.6E + 07	1.5	2.3E + 02	1	−
56	−0.9	4.2E + 07	1.8	2.9E + 03	55	1.9E − 02
57	−1.3	1.1E + 07	1.1	3.3E + 02	13	8.2E − 03
58	−1.4	8.1E + 06	1.4	5.3E + 01	2	−
59	−0.8	5.5E + 07	2.8	1.9E + 02	1	−
60	−0.7	7.7E + 07	1.7	2.5E + 03	1	−
61	−1.0	3.0E + 07	1.6	1.2E + 03	15	1.1E − 02
62	−1.3	1.0E + 07	1.4	2.4E + 02	23	6.7E − 03
63	−0.5	1.9E + 08	1.8	2.3E + 04	15	3.4E − 02
64	−1.3	1.1E + 07	1.2	1.0E + 02	2	−
65	−0.7	8.8E + 07	1.5	3.3E + 03	1	−
66	−1.6	4.6E + 06	1.2	2.5E + 01	3	−
67	−1.4	7.4E + 06	2.6	3.1E + 01	3	−
68	−1.0	3.5E + 07	1.5	2.7E + 03	65	2.1E − 02
69	−1.2	1.5E + 07	1.6	1.1E + 02	5	−
70	−1.0	3.7E + 07	1.0	1.9E + 03	2	−
71	−1.1	2.6E + 07	1.4	8.4E + 02	11	9.0E − 03
72	−0.9	4.9E + 07	1.5	9.0E + 02	1	−
73	−0.3	3.2E + 08	1.2	9.6E + 04	4	−
74	−0.8	7.3E + 07	1.4	2.4E + 03	1	−
75	−0.9	4.8E + 07	2.1	1.7E + 03	2	−
76	−1.3	1.3E + 07	1.3	1.5E + 02	3	−
77	−0.9	3.9E + 07	1.9	1.4E + 03	11	1.0E − 02
78	−0.7	1.0E + 08	1.6	5.5E + 03	5	−
79	−0.7	7.6E + 07	2.4	3.0E + 03	2	−
80	−0.6	1.2E + 08	1.7	5.6E + 03	1	−
81	−1.2	1.7E + 07	1.4	1.6E + 02	3	−
82	−0.9	4.1E + 07	1.6	1.5E + 03	12	1.0E − 02
83	−0.8	5.7E + 07	1.7	1.7E + 03	8	−
84	−0.5	1.9E + 08	1.4	2.1E + 04	3	−
85	−0.5	1.6E + 08	1.8	1.2E + 04	6	−

where $|\Omega_{0v}|$ represents the spectral level of the P wave and/or vector sum of the components of the S wave, ρ is the density of the source material (2700 kg/m^3), and F_v accounts for the radiation pattern of the P or S waves (as obtained from individual fault-plane solutions when available or average radiation coefficients of 0.52 and 0.63; BOORE and BOATWRIGHT, 1984). Magnitudes were obtained based on the moment magnitude relationship proposed by HANKS and KANAMORI (1979), $M = 2/3 \log M_0 - 6.0$, where M_0 is in N.m. The P- and S-wave seismic

Table 2

Source parameters as determined from spectral analysis for Site 2

No.	M	M_0 (N.m)	r_0 (m)	E_0 (J)	E_s/E_p	η
1	−2.4	2.7E + 05	1.3	1.6E − 01	25	−
2	−2.1	7.2E + 05	1.2	3.1E + 00	39	−
3	−1.6	4.1E + 06	1.9	9.7E + 00	17	−
4	−1.9	1.3E + 06	1.6	1.0E + 00	12	−
5	−1.7	2.6E + 06	2.0	2.4E + 00	9	−
6	−1.6	3.6E + 06	1.2	7.5E + 02	57	−
7	−1.7	3.3E + 06	1.4	5.3E + 01	54	6.7E − 04
8	−1.8	2.4E + 06	1.6	3.6E + 00	10	3.5E − 03
9	−1.2	1.7E + 07	1.5	3.9E + 02	28	−
10	−1.9	1.7E + 06	1.0	1.5E + 01	34	−
11	−1.7	3.3E + 06	1.9	1.4E + 01	54	2.7E − 03
12	−1.6	4.3E + 06	1.2	5.2E + 01	22	1.7E − 03
13	−1.6	4.6E + 06	1.7	2.8E + 01	37	4.1E − 03
14	−2.0	9.2E + 05	1.3	1.5E + 00	39	4.2E − 03
15	−2.2	6.1E + 05	0.9	3.0E + 00	32	5.3E − 04
16	−2.0	1.0E + 06	1.3	6.1E + 00	32	1.2E − 03
17	−1.7	2.7E + 06	1.3	9.6E + 00	14	−
18	−2.3	3.6E + 05	0.7	4.5E + 00	53	−
19	−1.8	2.3E + 06	1.5	7.7E + 00	17	3.8E − 03
20	−2.2	4.5E + 05	1.4	2.6E − 01	30	−
21	−1.8	2.4E + 06	1.3	9.1E + 00	24	8.9E − 04
22	−1.5	5.0E + 06	1.9	4.9E + 01	40	3.5E − 04
23	−1.6	4.4E + 06	1.6	1.9E + 01	19	3.9E − 03
24	−2.4	2.4E + 05	1.1	3.2E − 01	6	−
25	−1.5	7.0E + 06	2.7	1.4E + 01	11	−
26	−2.2	4.3E + 05	1.2	7.0E − 01	25	1.1E − 03
27	−1.4	1.0E + 07	1.8	8.2E + 01	10	1.1E − 02
28	−2.0	1.1E + 06	1.3	3.8E + 00	25	−
29	−1.6	4.6E + 06	1.9	1.4E + 01	22	5.6E − 03
30	−1.9	1.2E + 06	1.1	3.4E + 00	14	9.3E − 04
31	−1.8	1.8E + 06	2.5	1.0E + 00	20	1.5E − 03
32	−1.8	2.1E + 06	1.8	3.0E + 00	12	2.7E − 03
33	−2.1	7.3E + 05	0.9	5.1E + 00	24	−
34	−2.4	2.1E + 05	0.7	3.4E − 01	15	−
35	−2.1	6.8E + 05	1.4	7.3E − 01	16	−
36	−2.2	6.4E + 05	1.2	4.2E + 00	38	−
37	−1.7	3.2E + 06	1.9	2.1E + 01	15	4.8E − 04
38	−1.8	2.1E + 06	1.0	1.7E + 01	6	1.8E − 03
39	−1.4	8.4E + 06	1.4	1.7E + 02	28	1.2E − 02
40	−2.0	1.0E + 06	1.1	1.8E + 01	15	−
41	−2.2	5.3E + 05	1.0	5.5E + 00	81	−
42	−1.6	5.0E + 06	1.2	5.2E + 01	5	3.9E − 03
43	−1.8	2.0E + 06	1.2	2.7E + 01	11	−
44	−1.8	2.3E + 06	1.0	4.0E + 01	12	8.0E − 04
45	−1.5	5.8E + 06	1.1	2.4E + 02	18	−
46	−2.0	1.2E + 06	1.2	2.3E + 00	16	−
47	−1.9	1.9E + 06	1.4	6.7E + 00	18	−
48	−2.0	1.2E + 06	1.2	3.1E + 00	9	−

Table 2 (*Contd*)

No.	M	M_0 (N.m)	r_0 (m)	E_0 (J)	E_s/E_p	η
49	−1.6	3.4E + 06	1.5	1.7E + 01	10	–
50	−1.9	1.6E + 06	1.3	6.6E + 00	16	–
51	−1.5	5.1E + 06	1.5	2.8E + 01	15	7.4E − 03
52	−1.9	1.4E + 06	1.7	1.7E + 00	16	–
53	−2.0	1.4E + 06	1.2	1.2E + 01	29	–
54	−2.2	5.9E + 05	1.2	3.2E + 00	49	–
55	−1.7	3.5E + 06	2.2	8.6E + 00	17	–
56	−1.3	1.2E + 07	2.5	2.2E + 01	9	–
57	−1.6	4.4E + 06	1.1	2.8E + 02	62	–
58	−2.0	1.1E + 06	1.5	1.2E + 00	22	–
59	−2.1	6.8E + 05	1.1	8.4E − 01	8	–
60	−1.8	2.3E + 06	1.4	7.4E + 00	10	–
61	−1.9	1.6E + 06	1.1	6.4E + 00	15	–
62	−2.1	8.3E + 05	0.8	7.3E + 00	36	–
63	−2.1	7.3E + 05	0.9	4.8E + 00	23	–
64	−2.1	8.2E + 05	1.1	3.0E + 00	27	–
65	−2.1	7.8E + 05	1.0	7.7E + 00	47	–
66	−1.8	2.4E + 06	1.6	4.8E + 00	15	3.6E − 03
67	−2.0	1.1E + 06	1.6	8.3E − 01	18	–
68	−2.0	1.3E + 06	1.5	2.6E + 00	15	2.1E − 03

energies were calculated from the energy flux J_v following SNOKE (1987)

$$E_v = 4\pi\rho v R^2 J_v \langle F_v^2 \rangle / F_v^2 \tag{2}$$

where $\langle F_v^2 \rangle$ is the average radiation coefficient. Total seismic energy was calculated as $E_0 = E_p + E_s$. Estimates of source radius (r_0) were obtained by assuming MADARIAGA's (1976) model

$$r_0 = \frac{K_v \beta}{2\pi f_v} \tag{3}$$

where K_v is as obtained from Madariaga's graphical relationships or is equal to 2.01 and 1.32 for P and S waves, respectively, and f_v is the corner frequency

$$f_v = (J_v / 2\pi^3 \Omega_{0v}^2)^{1/3}. \tag{4}$$

If we assume that f_{max}, as obtained from the S wave, is related to the source, a radius r_a was obtained by replacing f_v with f_{max} in (3). The static stress drop $\Delta\sigma$ and the apparent stress σ_a were calculated as

$$\Delta\sigma = \frac{7}{16} \frac{M_0}{r_0^3} \tag{5}$$

$$\sigma_a = \frac{\mu E_0}{M_0} \tag{6}$$

where $\mu = \rho\beta^2$.

Table 3

Source parameters as determined from time domain analysis for Site 2. For available fault plane solutions, rupture velocities v_a and v_b correspond to each of the nodal planes as a plane of failure

No.	$\Delta\sigma_d$ (Pa)	σ_{rms} (Pa)	r/r_0	r_a/r_0	c	v_a ($\times\beta$)	v_b ($\times\beta$)
1	6.1e + 04	1.5e + 05	0.45	0.20	2.2	–	–
2	1.4e + 05	4.1e + 05	0.69	0.19	1.1	0.32	0.32
3	3.0e + 05	9.9e + 05	0.59	0.24	2.3	0.30	0.32
4	7.0e + 04	1.9e + 05	0.46	0.24	1.5	0.23	0.23
5	7.6e + 04	2.7e + 05	0.59	0.21	1.9	0.37	0.38
6	2.2e + 06	4.7e + 06	0.61	0.25	1.5	0.34	0.36
7	1.6e + 06	3.5e + 06	0.58	0.24	1.5	0.21	0.22
8	9.8e + 04	3.5e + 05	0.70	0.23	1.1	0.24	0.25
9	1.9e + 06	4.6e + 06	0.53	0.23	1.2	0.28	0.28
10	5.4e + 05	1.3e + 06	0.49	0.22	1.3	0.24	0.25
11	2.6e + 05	6.9e + 05	0.56	0.24	1.3	0.23	0.24
12	1.0e + 06	2.6e + 06	0.59	0.25	1.5	0.31	0.41
13	5.4e + 05	1.5e + 06	0.52	0.21	1.8	0.30	0.28
14	1.3e + 05	5.1e + 05	0.36	0.34	1.6	0.23	0.22
15	2.1e + 05	1.9e + 06	0.76	0.24	1.4	0.35	0.38
16	3.4e + 05	6.7e + 05	0.77	0.31	1.3	0.30	0.29
17	4.4e + 05	7.5e + 05	0.42	0.25	1.4	0.18	0.18
18	2.3e + 05	2.3e + 06	0.65	0.23	1.1	0.23	0.24
19	4.2e + 05	8.7e + 05	0.39	0.23	1.4	0.24	0.26
20	6.3e + 04	1.9e + 05	0.49	0.21	1.4	0.30	0.30
21	3.3e + 05	1.0e + 06	0.78	0.32	1.4	–	–
22	6.7e + 05	2.0e + 06	0.64	0.23	1.1	0.26	0.30
23	4.6e + 05	1.1e + 06	0.67	0.28	2.1	0.39	0.42
24	8.6e + 04	2.5e + 05	0.58	0.20	1.3	–	–
25	3.4e + 05	8.4e + 05	0.84	0.27	1.7	–	–
26	5.6e + 04	1.7e + 05	0.66	0.22	1.7	0.33	0.35
27	1.2e + 06	2.2e + 06	0.64	0.36	1.1	0.44	0.55
28	1.7e + 05	5.7e + 05	0.65	0.29	1.2	–	–
29	3.4e + 05	8.6e + 05	0.59	0.26	1.4	0.31	0.38
30	1.6e + 05	4.8e + 05	0.80	0.39	1.1	0.46	0.34
31	1.3e + 05	2.8e + 05	0.60	0.23	1.2	0.24	0.23
32	1.3e + 05	3.3e + 05	1.08	0.26	1.4	–	–
33	1.8e + 05	6.2e + 05	0.56	0.28	1.1	0.20	0.20
34	2.2e + 04	1.1e + 05	0.55	0.23	1.2	–	–
35	1.2e + 05	2.8e + 05	0.79	0.26	1.3	–	–
36	1.3e + 05	4.4e + 05	0.53	0.25	1.5	0.32	0.32
37	6.0e + 05	1.0e + 06	0.69	0.29	1.0	–	–
38	4.5e + 05	1.2e + 06	0.90	0.37	1.2	0.31	0.39
39	1.2e + 06	2.9e + 06	0.64	0.31	1.5	0.35	0.39
40	3.4e + 05	9.6e + 05	0.56	0.23	1.5	0.30	0.31
41	3.5e + 05	8.4e + 05	0.62	0.36	1.5	0.31	0.35
42	7.5e + 05	1.8e + 06	0.89	0.32	1.4	0.48	0.47
43	3.4e + 05	1.1e + 06	0.36	0.18	1.3	0.25	0.29
44	7.1e + 05	1.8e + 06	0.59	0.24	1.1	0.28	0.28
45	2.0e + 06	3.9e + 06	1.05	0.35	1.3	–	–
46	1.5e + 05	3.9e + 05	0.50	0.25	2.0	0.30	0.30
47	3.2e + 05	7.8e + 05	0.68	0.35	1.1	0.33	0.29

Table 3 (*Contd*)

No.	$\Delta\sigma_d$ (Pa)	σ_{rms} (Pa)	r/r_0	r_a/r_0	c	v_a ($\times\beta$)	v_b ($\times\beta$)
48	2.4e + 05	5.5e + 05	0.61	0.32	1.2	–	–
49	4.1e + 05	1.1e + 06	0.70	0.29	1.6	–	–
50	5.2e + 05	1.2e + 06	0.68	0.31	2.4	–	–
51	6.9e + 05	1.7e + 06	0.57	0.33	1.6	0.33	0.34
52	1.8e + 05	3.3e + 05	0.60	0.26	1.3	0.30	0.35
53	4.6e + 05	1.3e + 06	0.56	0.28	1.7	0.36	0.34
54	1.3e + 05	3.6e + 05	0.60	0.29	1.3	0.43	0.50
55	2.1e + 05	6.7e + 05	0.51	0.23	1.2	0.34	0.40
56	3.2e + 05	9.7e + 05	0.54	0.26	1.3	–	–
57	3.9e + 06	7.9e + 06	0.30	0.20	1.7	–	–
58	1.1e + 05	3.7e + 05	0.57	0.20	1.2	0.35	0.37
59	8.4e + 04	3.3e + 05	0.31	0.19	2.4	–	–
60	4.8e + 05	1.0e + 06	0.81	0.21	2.6	0.45	0.50
61	2.0e + 05	8.1e + 05	0.42	0.25	1.8	0.34	0.38
62	5.7e + 05	1.4e + 06	0.45	0.28	1.1	–	–
63	2.0e + 05	6.5e + 05	0.55	0.22	1.5	–	–
64	1.4e + 05	6.4e + 05	0.57	0.25	1.4	0.29	0.30
65	4.2e + 05	1.0e + 06	0.61	0.26	1.3	0.27	0.28
66	2.3e + 05	5.9e + 05	0.65	0.25	1.4	0.29	0.32
67	1.1e + 05	2.2e + 05	0.49	0.24	1.4	–	–
68	1.5e + 05	4.5e + 05	0.55	0.31	1.2	–	–

Supplementary source parameters were obtained from time domain analysis (Table 3). Peak acceleration $\underline{a}$ and velocity $\underline{v}$ values were used to determine the radius r of the most energetic asperity, which results in the maximum stress release (dynamic stress drop $\Delta\sigma_d$) according to the inhomogeneous model proposed by McGARR (1981). Assuming that the asperity can be considered as a Madariaga source within itself, then

$$r = 1.32\beta\underline{v}/\underline{a}$$
$$\Delta\sigma_d = 12.60\rho R\underline{a}. \tag{7}$$

An additional estimate of stress drop was calculated from measurements of the rms acceleration, a_{rms}, following the relationship originally established by HANKS and McGUIRE (1981) for a Brune source model (BRUNE, 1970), and extended here to the Madariaga model as

$$\sigma_{rms} = \frac{15.04\rho R}{F_v}\left(\frac{f_\beta}{f_{max}}\right)^{1/2} a_{rms} \tag{8}$$

where a_{rms} values are taken between the S-wave pulse initiation and $1/f_\beta$.

A measure of the source complexity was obtained by comparing the static stress drop with other estimates of stress release and by comparing the source radius with

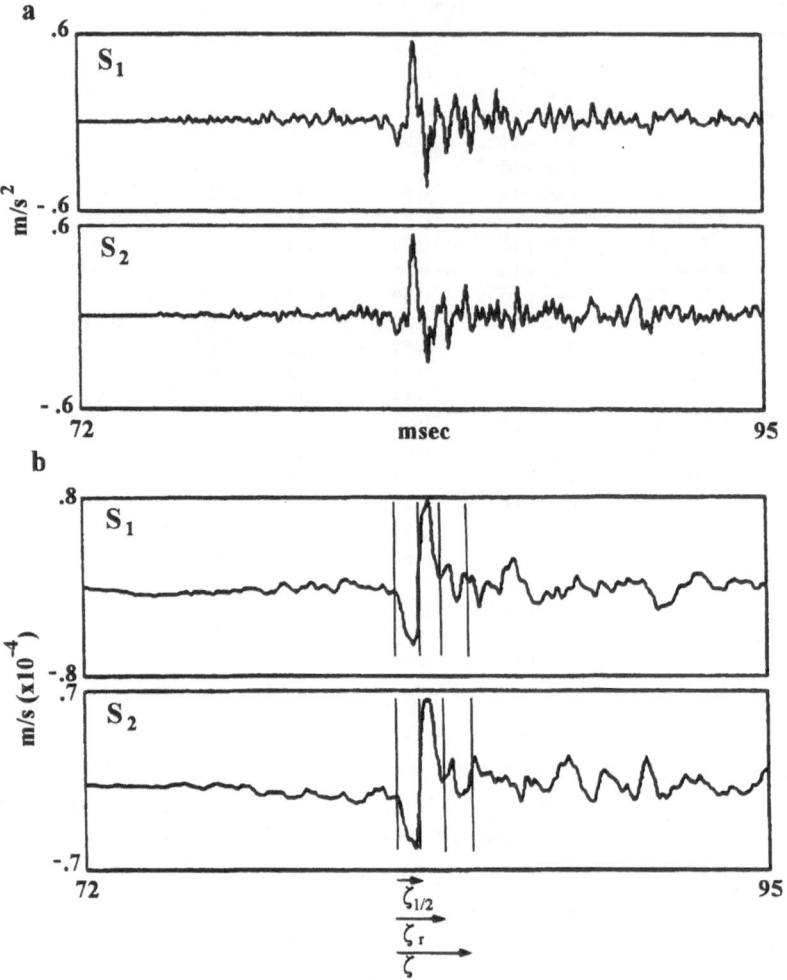

Figure 6
Example of rotated S-wave accelerations (a) and velocities (b), along with complexity measurements.

the asperity radius. Following the approach of BOATWRIGHT (1984), an independent estimate of source complexity was obtained by comparing the S-wave velocity signal duration of the slip over the entire source (ζ) to the pulse duration of the rupture of the first sub-event (ζ_r) as

$$c = \frac{\zeta/\zeta_r + \gamma}{1 + \gamma} \qquad (9)$$

where γ is the waveform compactness, taken to be between 0.5 and 1.0. Examples of signal and pulse duration measurements are given in Figure 6. Based on the measured velocity pulse rise time, $\zeta_{1/2}$, estimates of the rupture velocity v_r (BOATWRIGHT, 1980) were obtained as

$$v_r = \frac{13r}{16\zeta_{1/2} + 12r \sin \theta/\beta} \tag{10}$$

where θ is the angle between the fault normal and the takeoff direction as obtained from the fault-plane solutions.

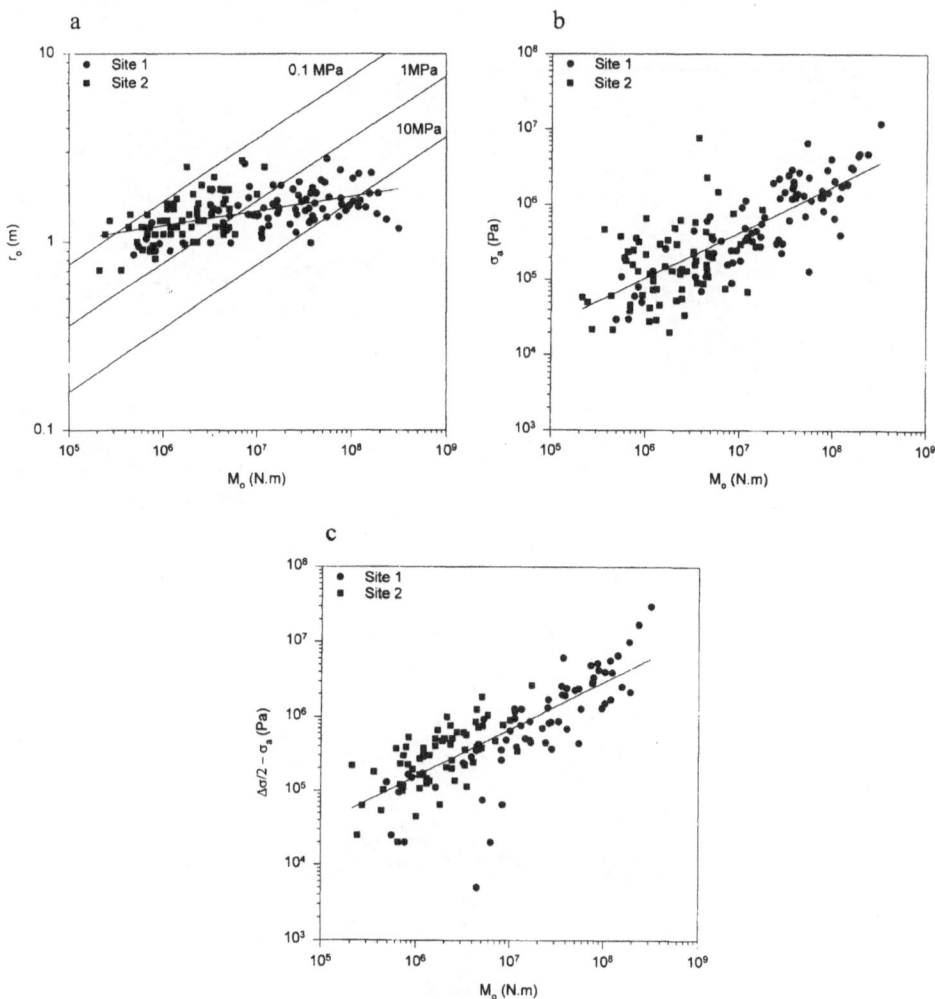

Figure 7
Scaling relationships: source radius (a), apparent stress (b), and excess stress release (c) as a function of seismic moment. To facilitate the visualization of trends, linear regression lines are shown. Additionally, constant static stress drop lines are plotted in (a).

Scaling

The scaling behaviour was examined for a combined data set consisting of events recorded at Sites 1 and 2. In Figure 7a, a clear departure from self-similar scaling is observed. For a constant stress drop ($\Delta\sigma$) of 1 MPa, it is expected that an increase in seismic moment from about 10^5 to 10^8 N.m would result in an increase in source radii (r_0) by a factor of 10, from about 0.3 to 3 m. Instead, the average source radii in our data remain relatively constant, with values of 1 to 2 m, and consequently,

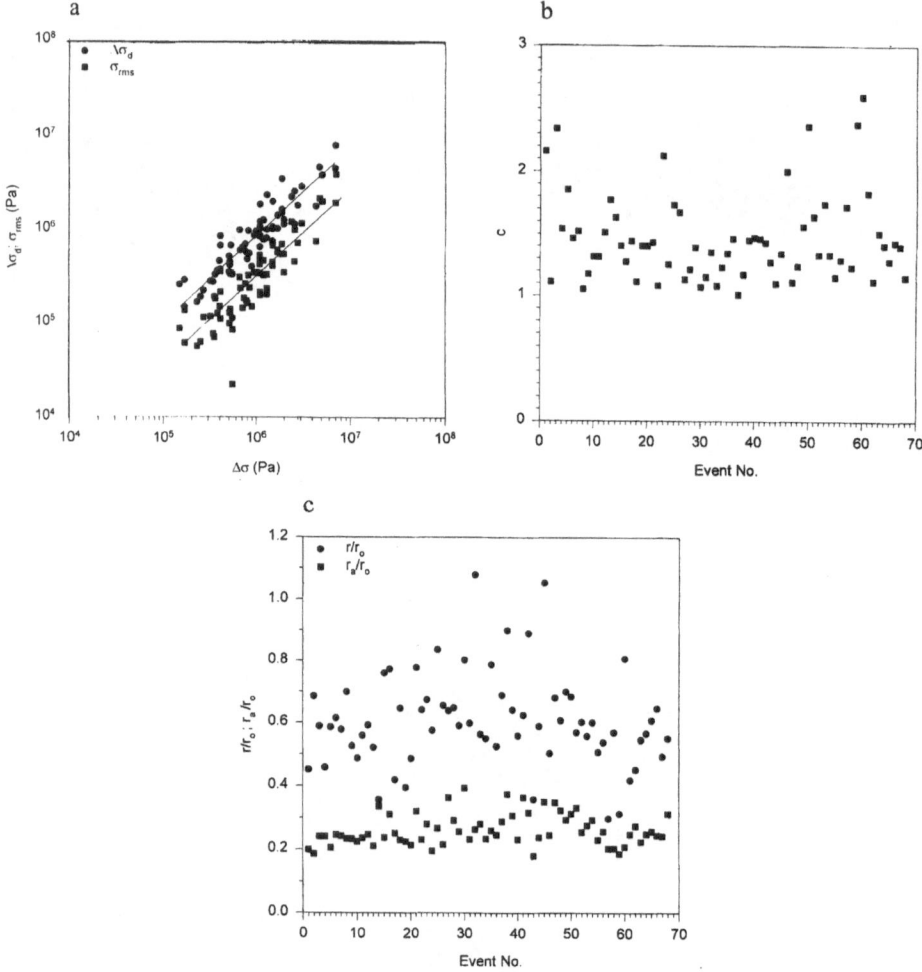

Figure 8
Geometric and rupture effects for Site 2: (a) comparison of stress release estimates, (b) source rupture complexity measurements, and (c) relationship of the most energetic asperity radius (circles) and minimum resolvable asperity radius (squares) to overall source radius.

the stress drops vary by a factor greater than 100, ranging from about 0.1 to more than 20 MPa. Additionally, the apparent stress (σ_a) also varies in average from about 0.03 to 3 MPa over the same seismic moment range (Fig. 7b).

Since $\Delta\sigma$ measures the difference between the initial (σ_1) and final stress (σ_2) levels on the fault, and σ_a measures the difference between the average stress ($\bar{\sigma}$) and average resisting stress ($\bar{\sigma}_r$), then the difference between $\Delta\sigma/2$ and σ_a (excess stress release) equals the difference between the average resisting and final stresses ($\Delta\sigma/2 - \sigma_a = \bar{\sigma}_r - \sigma_2$). From Figure 7c, $\Delta\sigma/2 - \sigma_a$ increases from about 0.5 to 7 MPa. This is possible if either $\bar{\sigma}_r$ increases and/or σ_2 decreases with increasing seismic moment. Moreover, since $\Delta\sigma$ and σ_a exhibit an increase with increasing seismic moment (Figs. 7a and 7b), then it is likely that σ_1 also increases.

The observations in Figure 7c are consistent with the concept of overshoot, which requires that the average resisting stress remains larger than the final stress. It may be considered that overshoot can occur when the rupture front encounters a high strength barrier under high average rupture velocities (MADARIAGA, 1976), or when healing takes place behind an advancing rupture front under low average rupture velocities. To examine these effects on scaling behaviour, in the following sections we investigate the role of complexity, in terms of source dimension and rupture velocity, and the role of shear and normal stress conditions at the time of failure.

Geometric and Rupture Effects

Geometric effects were examined by comparing stress release and source radius estimates as defined for homogenoeus and inhomogeneous source models. In terms of stress release, as shown in Figure 8a for Site 2, well-defined linear relationships were found between the $\Delta\sigma$ and dynamic stress drop ($\Delta\sigma_d$), and between $\Delta\sigma$ and rms stress drop (σ_{rms}). Moreover, $\Delta\sigma$ and $\Delta\sigma_d$ were similar in magnitude, whereas the σ_{rms} values were about 50% smaller. However, we know that the estimated errors in the determination of individual stress release values are at least 25 to 50%, suggesting that these stress release estimates are rather similar. Considering the estimates of source radii for Site 2 (Fig. 8b), the ratio between the most energetic asperity radius and overall source radius (r/r_0) ranges from 0.5 and 0.8, well above the minimum resolvable asperity level of 0.2 as suggested by the ratio of r_a/r_0. In Figure 8c, based on an independent measurement of source complexity made with (9), 75% of the events have c values below 1.5. Since c was introduced to characterize the possible occurrence of sub-events under an inhomogeneous stress distribution, the above c values suggest that the sources are geometrically relatively simple, as seen in both the $\Delta\sigma_d/\Delta\sigma$ and r/r_0 values.

As discussed in the previous section, the observed differences between the static stress drop and the apparent stress may be related to complex rupture propagation effects. Following RUDNICKI and KANAMORI (1981), the observed r/r_0 ratios in our

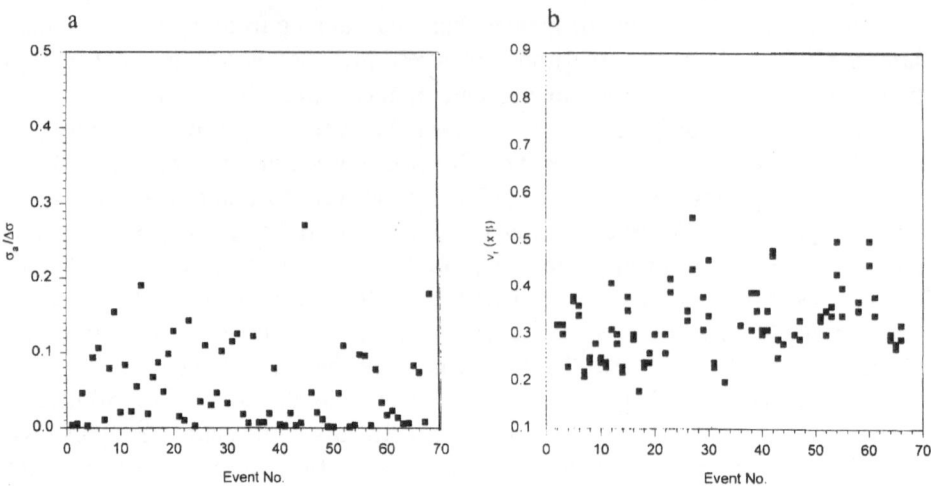

Figure 9
Site 2 rupture effects: (a) ratio of apparent stress to static stress drop, and (b) rupture velocity assuming that failure occurred on either of the two nodal planes.

study are unable to give values of $\sigma_a/\Delta\sigma$ below 0.05 (for Madariaga's model). However, 33% of the analyzed events for Site 2 had $\sigma_a/\Delta\sigma$ values below 0.03 (Fig. 9a). One possible explanation for this apparent complexity may reside in rupture velocities varying from 0.9β. Using (10), estimates of rupture velocity were found to range from 0.2 to 0.5β (both nodal planes were considered as faulting planes), considerably lower than the traditionally assigned value of 0.9β (Fig. 9b). When we considered r_0 in place of r in (10), the rupture velocities ranged from 0.3 to 0.6β. These low rupture velocities would allow for healing to occur behind the rupture front and thus create the conditions for overshoot as previously seen in Figure 7c.

Influence of Stress

In this study, three-dimensional boundary element numerical modelling was carried out for different excavation stages related to Site 2. Only clustered events belonging to the groups FW1, FW2, and C4 (see Fig. 3b), and not associated with the period of non-mining, were retained for this analysis. Maximum shear (τ) and normal stresses (σ_n) were calculated at every event location, along failure planes of infinite length with orientations as obtained from underground mapping. These stress values were then compared to derived failure criteria curves, based on geomechanical measurements, to isolate the most likely fracture set to fail. For example, the calculated maximum shear and normal stresses are shown in Figure 10 for the FW2 footwall events with respect to the failure curves derived for the three fracture sets considered from Figure 2b. Of the three sets, failure is only predicted for the second fracture set with an average strike/dip of 236°/60° (Fig. 10b).

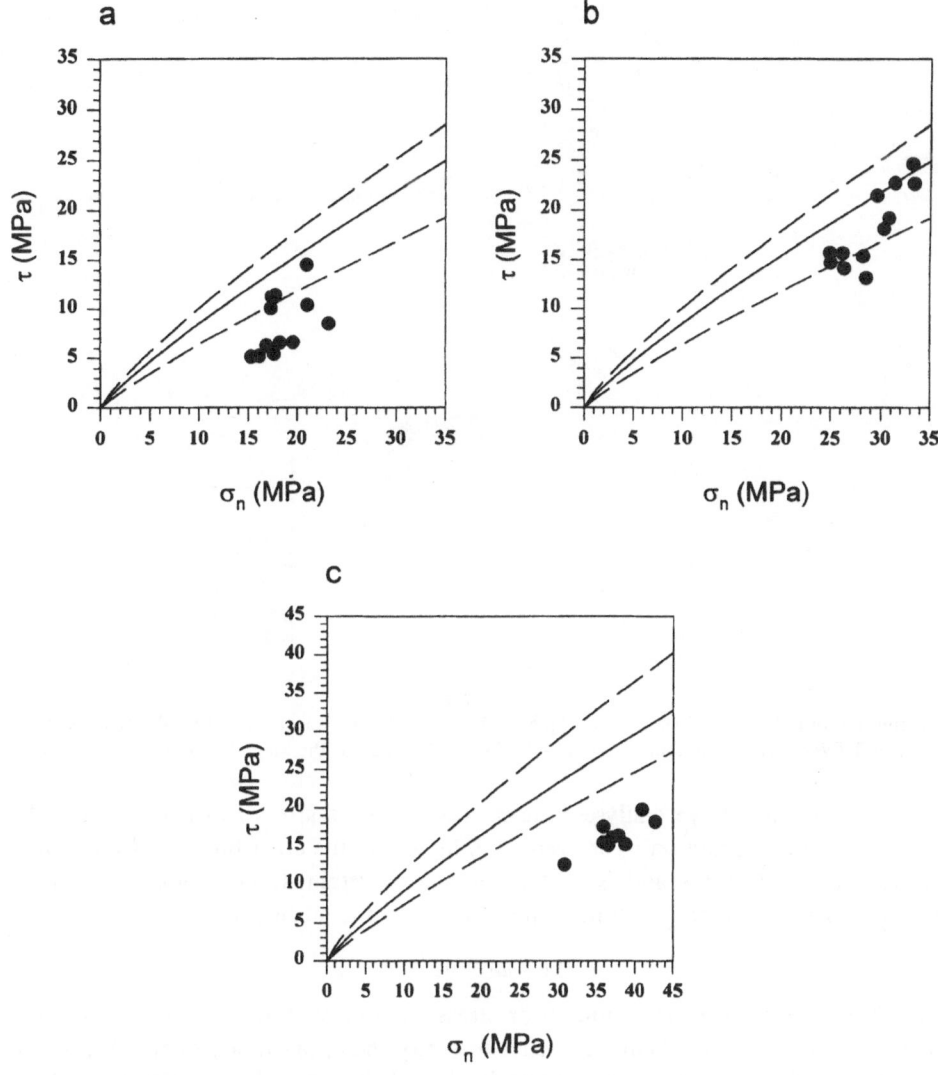

Figure 10

Maximum shear and normal stresses evaluated at FW2 event locations (Figs. 3c,d) for the three fracture sets identified in Figure 2b, respectively. Failure curves (continuous lines) and their error bounds (dashed lines), as determined from geomechanical measurements are also plotted. As seen, the calculated stresses predict failure only for fracture set 2 (b).

Interestingly, the highest concentration of poles to the most likely fault planes for these footwall events (Fig. 11), as determined from stress inversion of the fault-plane solutions, suggests that failure was mainly related to the second fracture set (240°/73°). Similar numerical modelling and stress inversion results were found for the FW1 and C4 groups of events.

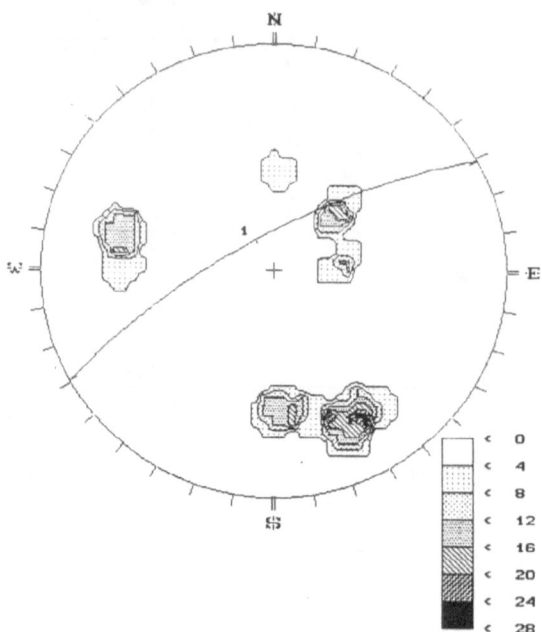

Figure 11
Distribution of poles to the identified fault planes, based on the stress inversion of focal mechanisms for
FW1 and FW2 events. The fault plane at 240°/73° corresponds to the highest concentration of poles.

Based on the above analyses, the maximum shear stress as calculated for the
second fracture set and only for events falling within the error bounds of the failure
criteria curve (BARTON and BANDIS, 1990), were retained. These maximum shear
stress values were used to estimate the Site 2 seismic efficiency as

$$\eta_{\text{model}} = \sigma_a/\tau. \tag{11}$$

For Site 1, values for maximum shear stress were only attainable through the use
of *in situ* stress measurements. In order to verify the applicability of this approach,
the seismic efficiencies were also determined for Site 2, based on *in situ* stress data,
$\eta_{\text{in situ}}$, and compared to η_{model}. The presence of an excavation, in general terms, has
a variable effect on the calculated maximum shear stress with distance from the
excavation. For locations close to the opening, this influence could be as large as
two times the measured far-field *in situ* stresses, whereas at distances equivalent to
about 1.5, excavation diameters result in a marginal influence on the calculated
shear stress values. For Site 2, the FW1 and FW2 events lie less than one diameter
from the opening. As a result, we calculated the maximum shear stress for these
events as

$$\tau_{\text{in situ}} = (2s_1 - s_3)/2 \tag{12}$$

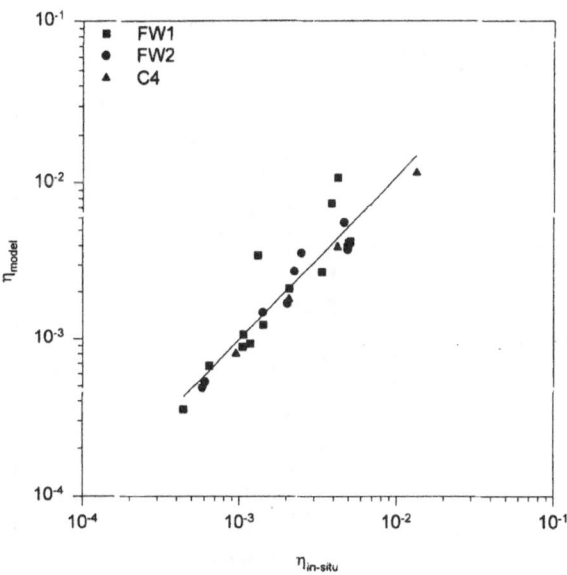

Figure 12
Comparison of seismic efficiencies based on numerical modelling derived maximum shear stresses (η_{model}) with those computed using *in situ* stress measurements ($\eta_{in\ situ}$). The line corresponds to a one-to-one relationship.

where s_1 and s_3 are the maximum and minimum principal stress magnitudes. The C4 events were situated somewhat further from the opening and a factor of 1.5 was used in place of 2 for s_1. As shown in Figure 12, a good correlation exists between η_{model} and $\eta_{in\ situ}$ for Site 2, suggesting that *in situ* measurements, corrected for distance from the opening, can be used to determine the maximum shear stress and consequently the seismic efficiencies for Site 1.

The events analyzed for Site 2 had S-wave to P-wave energy ratios greater than ten, whereas a number of events from Site 1 were characterized by ratios less than ten. For the purpose of consistency, only events with energy ratios greater than ten were retained for further analysis. Figure 13 shows the seismic efficiencies as a function of seismic moment for the two sites as obtained using (12) and (11), respectively (Tables 1 and 2). The seismic efficiencies ranged from about 0.0003 to 0.06, similar to values reported by McGarr (1994). In spite of the observed scattering, a clear, positive correlation of seismic efficiency with seismic moment is found. This scaling parallels the behaviour previously observed in the apparent stress (see Fig. 7b), indicating that seismic stress release estimates are quasi-independent of the maximum shear stress level and that other stress parameters, such as σ_n or stress invariants (Kagan, 1994), may play an important role in the failure process.

In order to examine the stress conditions leading up to failure, τ and σ_n were calculated for the second fracture set at locations of seismicity during several

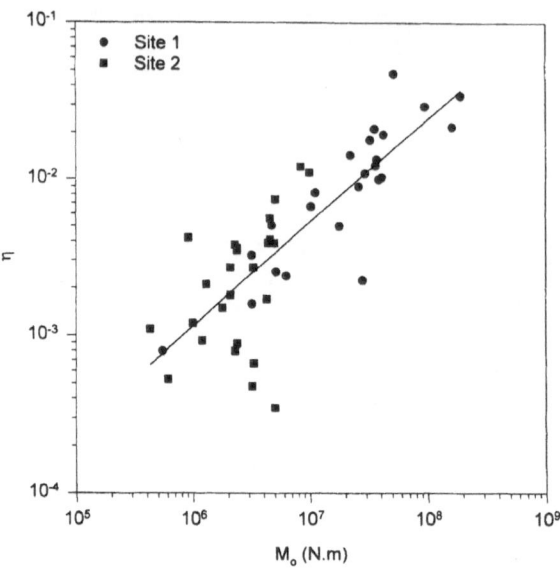

Figure 13
Seismic efficiency as a function of seismic moment. To visualize the trend, the best fit regression line is also shown.

progressive stages of excavation (Site 2). As shown in Figure 14, stress paths were determined for regions corresponding to FW1, FW2, and C4. For FW1 and FW2 (Figs. 14a, b), stress values were obtained for event locations prior to the excavation, and following the first and second stages of extraction, with the final point corresponding to the observed seismic event. In both cases, failure occurred by an increase in both shear and normal stresses as indicated by the proximity of the final points to the failure curve. In contrast, for C4, failure occurred when the normal stress decreased while the shear stress remained constant (stage three). These results emphasize the importance of σ_n on the failure process. Maintaining high levels of σ_n at failure, as seen for FW1 and FW2, would result in high average resisting stress levels. Under low average rupture velocities this may be responsible for the observed overshoot in Figure 7c.

Conclusions

This study has investigated the effects of source rupture complexity and stress regime on observed source parameter scaling for induced microseismic events ($-2.4 \leq M \leq -0.3$) at Strathcona mine, Sudbury, Ontario. Two excavation sites were considered, one associated with a region of elevated stress levels and a complex distribution of fracture orientations, and one associated with moderate

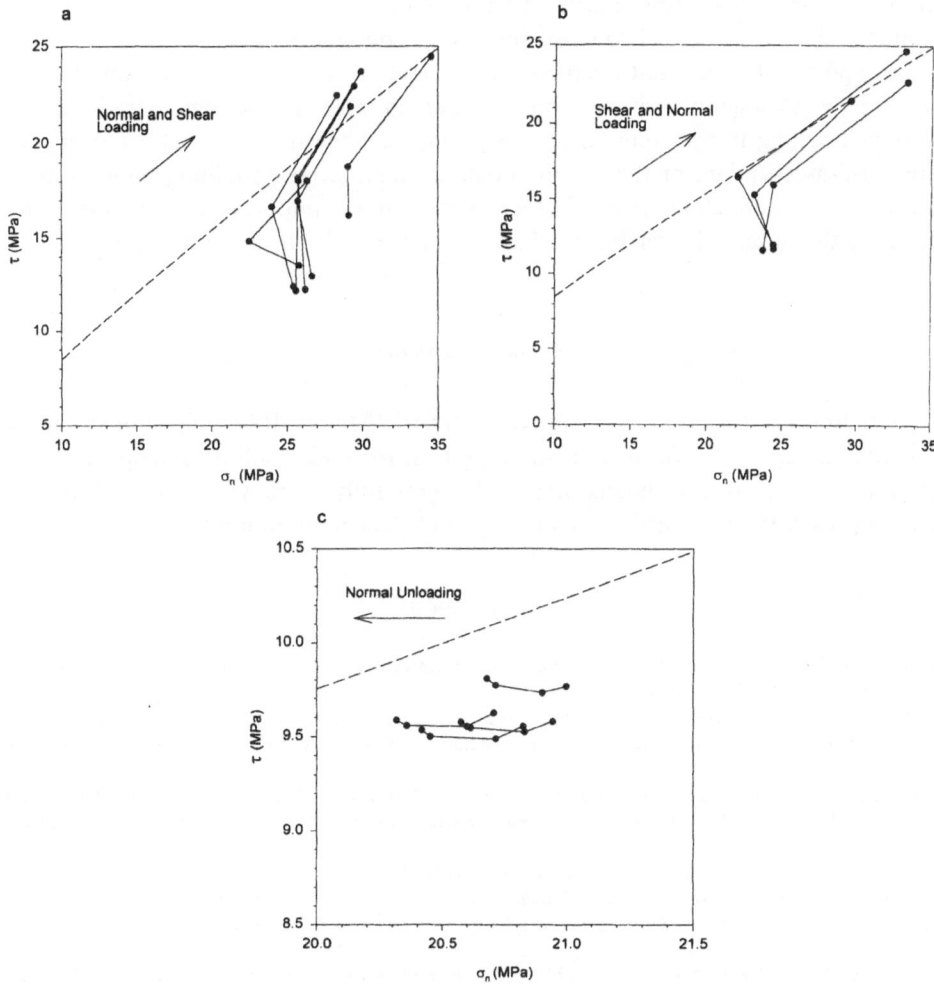

Figure 14
Shear and normal stresses as determined for progressive excavation stages, assuming that only fracture set 2 was active: (a) FW1 events, (b) FW2 events, and (c) C4 events.

stresses and rather simple fracturing. The analyzed seismic events were recorded with two underground microseismic networks. Scaling of source radii, apparent stress, and excess stress release as a function of seismic moment was derived from spectral analysis. Insight into the observed scaling was provided by examining waveform complexities, calculated rupture velocities, and maximum shear stresses.

A non-similar scaling behaviour was emphasized in our results, indicating that a relatively strong dependence existed between stress release, over 2 orders of magnitude, and seismic moment, over 3 orders of magnitude. The positive scaling in excess stress release was consistent with the concept of overshoot. Observed low rupture velocities, from 0.2 to 0.5β, and simple sources, with complexities less than 1.5, supported the idea that overshoot was related to the process of healing behind the slowly advancing rupture front. Scaling in seismic effciency paralleled the observed scaling in apparent stress, suggesting that seismic stress release estimates are quasi-independent of the maximum shear stress level. At failure, high levels of normal stress were emphasized, further supporting the importance of high resisting stress in the observed overshoot behaviour and its role in the failure process.

Acknowledgments

Funding for this study was provided by the Mining Research Directorate of Canada. We are thankful to Falconbridge Ltd. for their continued support of our research efforts at Strathcona mine. We gratefully acknowledge Sue Bird for assisting us with the numerical modelling and data preparation.

REFERENCES

AKI, K. (1984), *Asperities, Barriers, Characteristic Earthquakes and Strong Motion Prediction*, J. Geophys. Res. *89*, 5867–5872.

BARTON, N., and BANDIS, S. C., *Review of predictive capabilities of JRC-JCS model in engineering practice.* In *Rock Joint* (eds. Barton, N., and Bandis, S. C.) (A. A. Balkema, Rotterdam 1990) pp. 603–610.

BIRD, S. (1993), *Link of Structural Mapping, Numerical Modelling, and Microseismic Source Parameters with Application to Mine Design*, M.Sc. Thesis, Department of Mining Engineering, Queen's University, Kingston, Canada.

BOATWRIGHT, J. (1980), *A Spectral Theory for Circular Seismic Sources; Simple Estimates of Source Dimension, Dynamic Stress-drop, and Radiated Seismic Energy*, Bull. Seismol. Soc. Am. *70*, 1–27.

BOATWRIGHT, J. (1984), *The Effect of Rupture Complexity on Estimates of Source Size*, J. Geophys. Res. *89*, 1132–1146.

BOORE, D. M., and BOATWRIGHT, J. (1984), *Average Body-wave Radiation Coefficients*, Bull. Seismol. Soc. Am. *74*, 1615–1621.

BRUNE, J. N. (1970), *Tectonic Stress and the Spectra of Seismic Shear Waves from Earthquakes*, J. Geophys. Res. *75*, 4997–5009. Correction in J. Geophys. Res. *76*, 5002 (1971).

CICHOWICZ, A., GREEN, R. W. E., BRINK, A. V. Z., GROBLER, P., and MONTFORT, P. I., *The space and time variation of microevents occurring in front of an active slope.* In *Rockbursts and Seismicity in Mines* (ed. Fairhurst, C.) (A. A. Balkema, Rotterdam 1990) pp. 171–175.

COCHRANE, L. B. (1991), *Analysis of the Structural and the Tectonic Environments Associated with Rock Mass Failures in the Mines of the Sudbury District*, Ph.D. Thesis, Department of Geological Sciences, Queen's University, Kingston, Canada.

FEUSTEL, A., URBANCIC, T. I., and YOUNG, R. P., *Estimates of Q using the spectral decay technique for seismic events with $M < -1$.* In *Rockbursts and Seismicity in Mines* (ed. Young, R. P.) (A. A. Balkema Rotterdam 1993) pp. 337–342.

FLETCHER, J. B., HAAR, L. C., VERNON, F. L., BRUNE, J. N., HANKS, T. C., and BERGER, J., *The effect of attenuation on the scaling of source parameters for earthquakes at Anza, California.* In *Earthquake Source Mechanics* (eds. Das, S., Boatwright, J., and Scholz, C. H.) Am. Geophys. Union, Washington, D.C. 1986) pp. 331–338.

GEPHART, J. W., and FORSYTH, D. W. (1984), *An Improved Method for Determining the Original Stress Tensor Using Earthquake Focal Mechanism Data: Application to the San Fernando Earthquake Sequence*, J. Geophys. Res. *89*, 9305–9320.
GIBOWICZ, S. J., YOUNG, R. P., TALEBI, S., and RAWLENCE, D. J. (1991), *Source Parameters of Seismic Events at the Underground Research Laboratory in Manitoba, Canada: Scaling Relations for Events with Moment Magnitude Smaller than* − 2, Bull. Seismol. Soc. Am. *81*, 1157–1182.
HANKS, T. C. (1977), *Earthquake Stress Drops, Ambient Tectonic Stress and Stresses that Drive Plate Motions*, Pure and Appl. Geophys. *115*, 441–448.
HANKS, T. C., and KANAMORI, H. (1979), *A Moment Magnitude Scale*, J. Geophys. Res. *84*, 2348–2350.
HANKS, T. C., and McGUIRE (1981), *The Character of High Frequency Strong Ground Motion*, Bull. Seismol. Soc. Am. *71*, 2071–2096.
KAGAN, Y. Y. (1994), *Incremental Stress and Earthquakes*, Geophys. J. Int. *117*, 345–364.
MADARIAGA, R. (1976), *Dynamics of an Expanding Circular Fault*, Bull. Seismol. Soc. Am. *66*, 639–666.
McGARR, A. (1981), *Analysis of Peak Ground Motion in Terms of a Model of Inhomogeneous Faulting*, J. Geophys. Res. *86*, 3901–3912.
McGARR, A. (1984), *Scaling of Ground Motion Parameters, State of Stress, and Focal Depth*, J. Geophys. Res. *89*, 6969–6979.
McGARR, A. (1994), *Some Comparisons between Mining Induced and Laboratory Earthquakes*, Pure and Appl. Geophys. *142*, 467–489.
NALDRETT, A. L., and KULLERUD, G. (1967), *A Study of the Strathcona Mine and its Bearing on the Origin of the Nickel-copper Ores of the Sudbury District, Ontario*, J. Petrology *8*, 453–531.
SNOKE, J. A. (1987), *Stable Determination of (Brune) Stress Drops*, Bull. Seismol. Soc. Am. *77*, 530–538.
RUDNICKI, J. W., and KANAMORI, H. (1981), *Effects of Fault Interaction on Moment, Stress Drop, and Strain Energy Release*, J. Geophys. Res. *86*, 1785-1793.
URBANCIC, T. I., FEIGNIER, B., and YOUNG, R. P. (1992), *Influence of Source Region Properties on Scaling Relations for M < 0 Events*, Pure and Appl. Geophys. *139*, 721–739.

(Received January 30, 1995, revised October 18, 1995, accepted October 24, 1995)

PAGEOPH, Vol. 147, No. 2 (1996)

0033–4553/96/020345–11$1.50 + 0.20/0

Simulation of Triggered Earthquakes in the Laboratory

G. A. SOBOLEV,[1] A. V. PONOMAREV,[1] A. V. KOLTSOV,[1] and V. B. SMIRNOV[1]

Abstract—The experiments were conducted for the study of stick-slip at the contact between two granite blocks. Three cases were studied under the following conditions: 1) the increase of load at a constant rate; 2) the additional application of sinusoidal oscillations in the frequency range from 1 to 30 Hz; 3) subjection to the impulse in the kilohertz frequency range. The imposition of sinusoidal oscillations with the amplitude of 15% of the maximal load caused the reduction of time by 10% for the next stick-slip occurrence, as compared to that expected during the gradual loading. This effect is discussed in terms of durability. The high frequency impulse influence increased this effect and also caused essential changes in the amplitude of elastic oscillations generating during the stick-slip. The trigger phenomena should be integrated in prediction models of the time and magnitude of earthquakes.

Key words: Stick-slip, rock, instability, model, earthquake.

Introduction

In the seismically active regions of the earth, the tectonic stresses do not change smoothly but fluctuate as a result of earth tides, under the influence of near earthquakes, technogenic explosions and meteorological factors. These phenomena may initiate movements along the existing faults, including stick-slip events that cause earthquakes. The latest example was the sudden increase of seismicity spanning a large area in California after the Landers earthquake in 1992 (BODIN and GOMBERG, 1994). The trigger effects are not included in the models of prediction of the time or magnitude of an earthquake based on the magnitude of the fault movement during the previous event in the same region; the rate of stress (strain) accumulation in such models is assumed constant (SHIMAZAKI and NAKATA, 1980).

Earlier laboratory experiments have shown (SOBOLEV *et al.*, 1991, 1993) that the influence of sinusoidal or impulse oscillations produce the reduction of the critical load threshold when a stick-slip occurs. The subjection to impulse caused the effect of greater elastic oscillation (microearthquake) to generate during the stick-slip

[1] Institute of Seismology, Russian Academy of Sciences, B. Grusinskaya 10, Moscow, Russia.

and supplemented the spectrum with high frequencies. This effect is not qualitatively dependent on the type of rocks and the size of the model.

In this paper, we endeavoured to quantitatively estimate the trigger effect of sinusoidal vibration by changing their frequency, amplitude, and duration.

The Methods Used in the Experiments

All experiments were conducted under conditions of the double-axes compression on a model composed of two granite blocks (see Fig. 1). The contact surface was 240 mm long and 70 mm wide; the inclination angle to the major (vertical) load was 25°. The friction surfaces were treated with the grid No. 1000. The structure of the model allowed the size of the contact area to remain constant during progressive movements of the upper block over the lower one. The lateral load G in this scheme was also nearly constant; it changed during stick-slip by no more than 1%. The steel liners with rollers were used to reduce the tangential stresses between the facets of the model and the pistons of the press.

The experiments were carried out on the servo-controlled press "Inova" at the Borok Observatory of the United Institute of Physics of the Earth. In all, 38

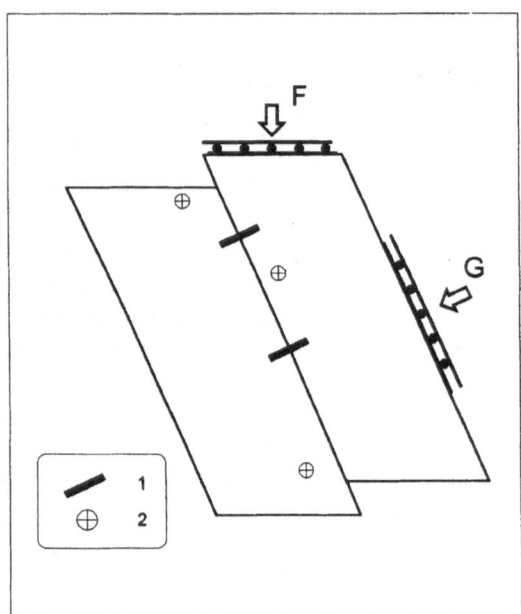

Figure 1
Loading arrangement: gouges of relative movements (1); PZT transducers (2).

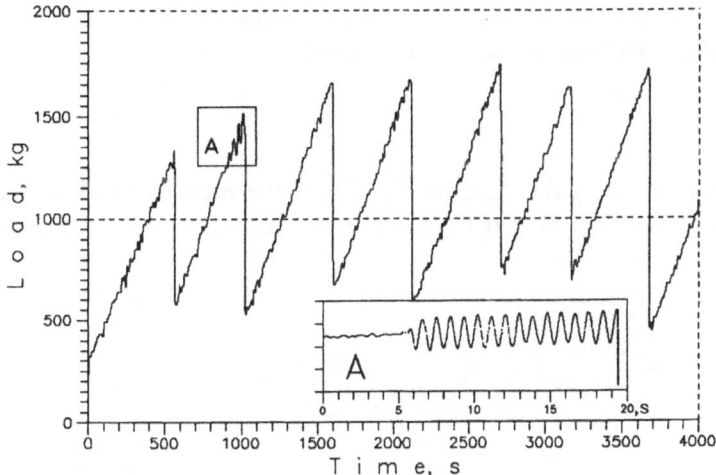

Figure 2
The example of stick-slip series.

experiments were accomplished; each of them contained from 8 to 12 subsequently occurring stick-slip events. The relative displacement of blocks along the contact during one event was 1 mm on the average. After each experiment, the model was dismantled for cleaning of surfaces, and the blocks were placed in the initial position.

The values of the major vertical load F and of the distance between the lower and upper pistons of the press were recorded with 1 s discretion on an IBM-PC. Two recorders of reciprocal movements of the edges of blocks d were glued along the contact; the recorders were composed of springs with strain gages. The data on movement d were recorded on the IBM-PC with a discretion interval of 10 ms (low frequency channel) and of 10 mcs (high frequency channel). Moreover, three piezoceramic receivers of the piston type 20 mm in diameter with fundamental frequency 0.1 MHz were fixed along the contact. During the stick-slip events, the receivers recorded the generation of elastic oscillations (microearthquakes) on the IBM-PC with a sampling rate from 2 to 10 mcs.

The experiments without additional vibration were conducted at a constant rate of strain equal to $5 \cdot 10^{-6} \, \text{s}^{-1}$. They alternated with the experiments that acquired the sinusoidal component, supplementing the linearly growing deformation. This arrangement resulted in a series of stick-slips which occurred after a period of several minutes (see Fig. 2). The following vibration frequencies were used: 1, 5, 10 and 30 Hz. The duration of the period of application of vibration varied. In some experiments the vibration was applied by short (1 s) impulses every 5 or 10 s. The amplitude of sinusoidal vibration was 15% in relation to the critical level under which the stick-slip occurred. Occasionally the experimentator initiated the stick-slip event with a shock on the surface of the model. In such cases the spectrum of the initiating pulse was in the kilohertz frequency range.

Results

Let us estimate in this kind of experiment the practicality of a model for the prediction of the time or magnitude of an earthquake (SHIMAZAKI and NAKATA,

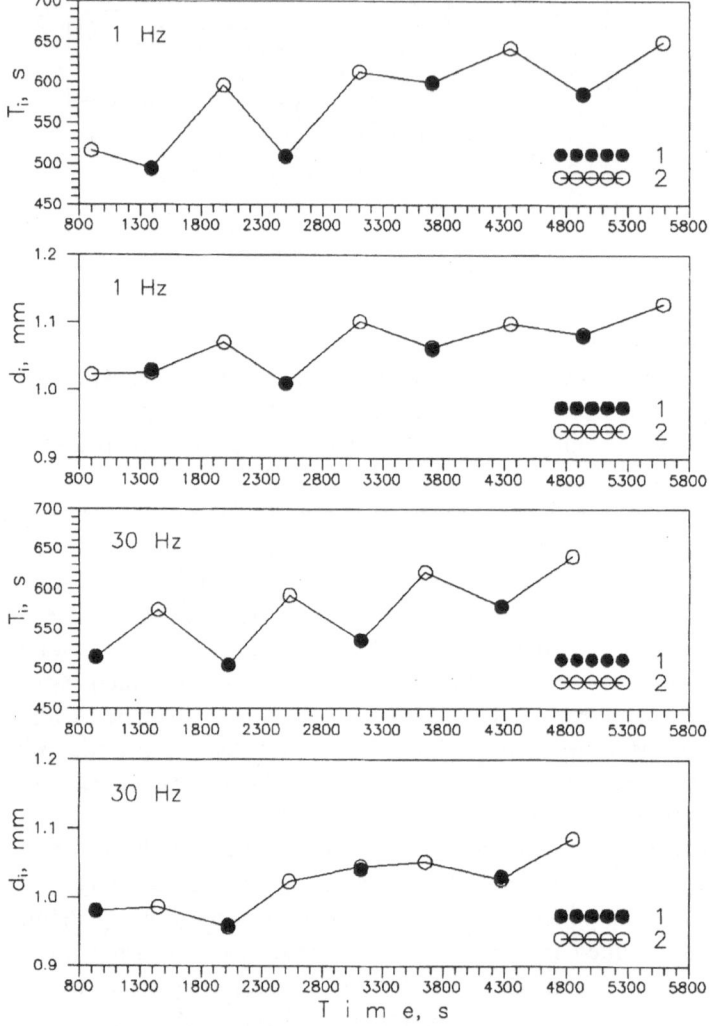

Figure 3
The variations of times T_i and relative displacements d_i, which took place under alternating loading with additional vibration (1) and smooth loading (2).

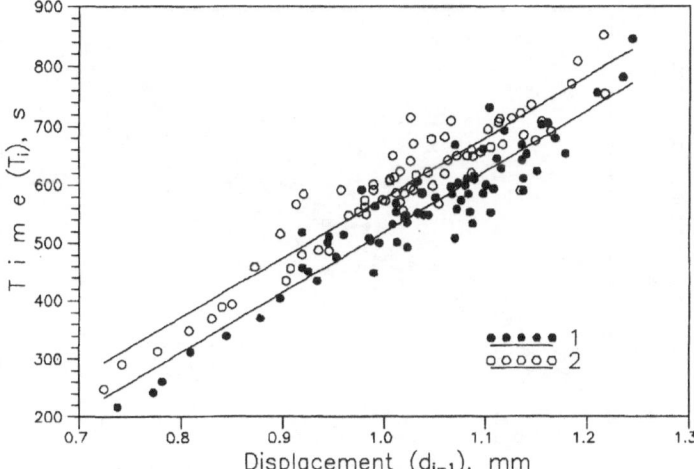

Figure 4

Graphs of regression $T_i = f(d_{i-1})$ for the sequence: smooth loading—additional vibration (1) and additional vibration—smooth loading (2).

1980). They presumed that at a constant rate of stress (deformation) accumulation on a certain fault, the time span until the next earthquake or its magnitude linearly depend on the extent of the relative movement along the fault during the previous earthquake. In the simple model of recurring stick-slip events on the contact between two blocks, this approach results in the following rules. If the recurring unstable movements appear at one and the same critical stress σ_{cr}, then the time—predictable model is valid:

$$T_i = Ad_{i-1} + B, \qquad (1)$$

where T_i is the time after the previous movement with magnitude d_{i-1}. Moreover, if during each movement the stress drops to the constant level of residual stress σ_r, then the magnitude of the next event shall be in proportion to time T_i and, consequently

$$d_i = Cd_{i-1} + D. \qquad (2)$$

In our case, each experiment generated from 8 to 12 recurring stick-slip events, which took place under the alternating smooth loading and the loading with superimposed additional vibration, whereby the stability of successive values of T_i and d_i could be controlled. During the data processing, the results of the first two stick-slip events were disregarded, owing to a sharp increase of critical stress indicating the attrition of the contact after the dismantling of the model. Figure 3 shows, as a typical example, the results of two experiments, when vibration was

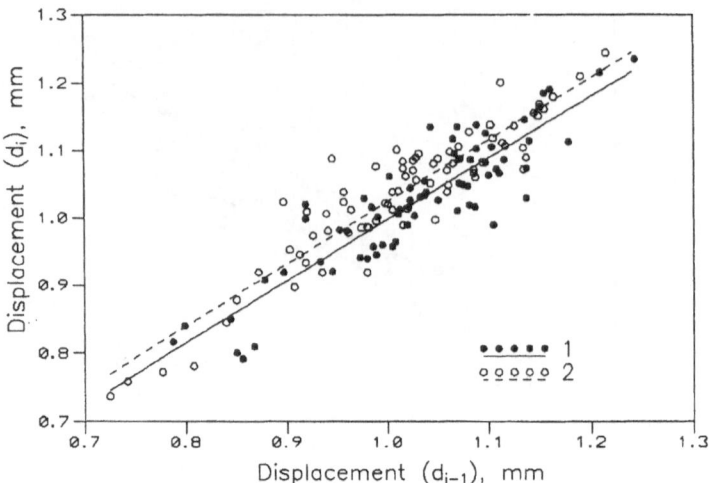

Figure 5
Graphs of regression $d_i = f(d_{i-1})$ for the sequence: smooth loading—additional vibration (1) and additional vibration—smooth loading (2).

applied at the lowest of the used frequencies, i.e., 1 Hz, and at the highest, i.e., 30 Hz. Similar results were obtained by vibration with the frequency 5 and 10 Hz.

Figure 3 demonstrates that the values T_i and d_i do not remain constant. The calculation of the regression $T_i = f(d_{i-1})$ produces the different results for the sequences: a) smooth loading—additional vibration; b) additional vibration—smooth loading. The corresponding relations

$$T_i^{av} = Ad_{i-1}^{sl} + B \quad \text{(for case a)} \tag{3}$$

$$T_i^{sl} = Cd_{i-1}^{av} + D \quad \text{(for case b)}, \tag{4}$$

where $A = 1037$ s/mm, $B = -518$, $C = 1028$ s/mm, $D = -450$ are shown in Figure 4. The sequence (3) produced the average reduction of T_i by about 10% in comparison with (4). The relations between the magnitudes of stick-slip events for cases (3) and (4) satisfy the equations

$$d_i^{av} = Kd_{i-1}^{sl} + L \quad \text{(for case a)} \tag{5}$$

$$d_i^{sl} = Md_{i-1}^{av} + N \quad \text{(for case b)}, \tag{6}$$

where $K = 0.92$, $L = 0.10$, $M = 0.91$, $N = 0.08$, and are illustrated in Figure 5. The sequence (5) produced the average reduction of d_i by about 2%.

The results described above were obtained at vibration amplitude 15% comprised by critical force, F_{cr}. In experiments with lower vibration amplitude (5% of F_{cr}) no significant difference between cases with additional vibration and without one was

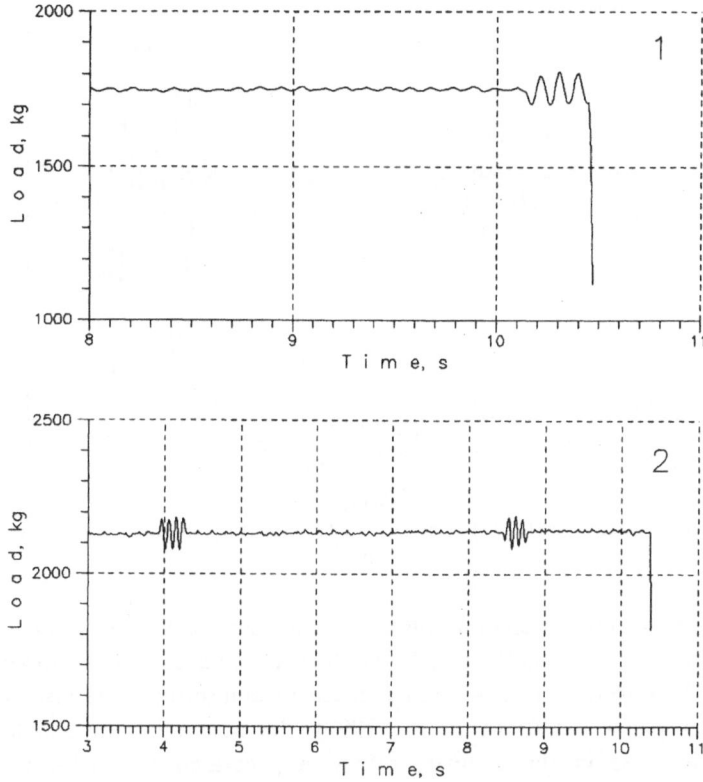

Figure 6

The examples of time delay between the moments of unstable movements triggered by vibration and peak stresses: 1) continual vibration, 2) short impulses.

obtained. At this stage of the studies we could not discern significant changes in T_i and d_i to be dependent on the frequency of vibration and the duration of its application.

At the same time, however, the effect of a delay of the stick-slip with regard to the beginning of initiation is clearly apparent. The typical examples shown in Figure 6, indicate that unstable movement started a few seconds after the stress reached its maximum. This delay occurred both in the case of vibration applied by impulses and under continuous action. It is noted, that the moment of stick-slip event occurs at a lower stress level as compared to that of previous local maximum(a). The number of these maxima diminished with the reduction of vibration frequency. In these experiments the longest delays between the first excess of the critical stress and the moments of corresponding stick-slip events were several seconds.

In a few cases, the stick-slip was initiated by a shock on the surface of the model, the energy bearing frequencies of which were in the range of several kHz,

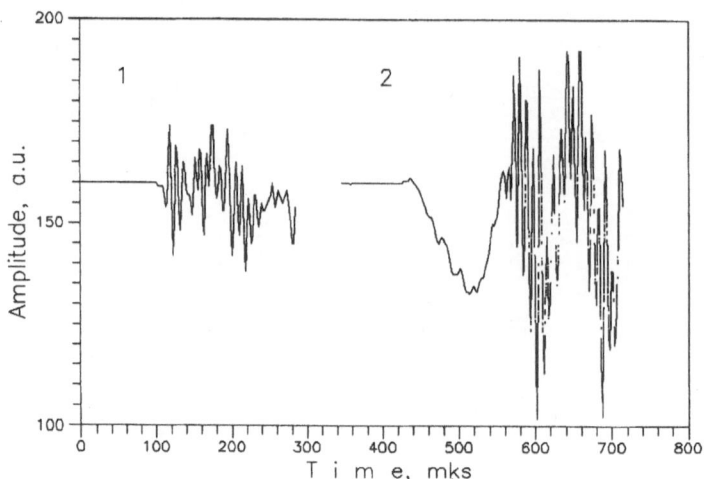

Figure 7
Acoustic radiation during stick-slip: (1)—initiated by low-frequency vibration, (2)—by high-frequency impulse.

i.e., thousands of times exceeding the frequencies applied by the pressed sinusoidal vibration. Even in this case the instability started with a delay from the moment of shock, but the duration of delay was reduced to milliseconds. As also represented in an earlier paper (SOBOLEV *et al.*, 1993), the action by the shock caused a significant increase in the acoustic radiation generated by stick-slip. Figure 7 demonstrates for comparison the seismograms during stick-slip without a shock (1), and caused by the shock (2). Figure 7(2) shows that 120 microseconds after the beginning of relatively low-frequency oscillation, caused by the shock, the high frequency response begins, and its amplitude is greater by an order than a similar amplitude in Figure 7(1). This effect is not observed in cases when the stick-slip was initiated by continuous or impulsive low-frequency (1–30 Hz) vibration.

Discussion

We shall discuss three possible mechanisms of the earlier beginning of the unstable movement in the experiments with vibration, in other words, the reduction of T_i (Figure 4). The reduction of d_i may be attributed to a corresponding decrease of critical stress.

One of the causes is probably the change of conditions on the contact during vibration, in particular, the reduction of the static coefficient of friction, which we estimate at $\mu = 0.6$. Accordingly when the edges of the contacted blocks are relatively displaced by vibration, μ can be reduced owing to the shorter time of the static contact or to the greater velocity of the relative moment, as demonstrated in the experiments (DIETERICH, 1979). Certain facts contradict this supposition. First, the influence of the frequency of vibration on the reduction value of T_i is not recorded. Second, the change in the dynamic pattern of acoustic radiation is not registered during the stick-slip with and without vibration, which is contrary to the case in which the stick-slip is initiated by high-frequency shock. Let us recall that according to the results given in SOBOLEV et al. (1993). The powerful radiation during the movement initiated by a shock was attributed to the absence of a previous creep along the contact and, as a consequence, to a higher roughness of the contact. Third, we found no significant difference of stress drop to the critical stress $\Delta\sigma/\sigma$ ratio in the cases described above.

Another possible reason for T_i and d_i reduction can be attributed to the effect of the durability. The effect of time dependency is clearly seen in Figure 6. Let this time—longevity—under constant applied stress σ be defined by an equation of kinetic strength theory

$$\theta = \Theta \, e^{\,U_0 - \gamma\sigma/RT_0},$$ (7)

where $\Theta \approx 10^{-13}$ s, $R = 2 \cdot 10^{-3}$ kcal/mol, T_0—absolute temperature, U_0, γ—activation energy and structure-sensitive parameters, estimated empirically. The longevity τ under varying stress can be found from condition (REGEL et al., 1974):

$$\int_0^{\tau} \frac{dt}{\theta(\sigma(t))} = 1.$$ (8)

Substituting in (8) the load regimes, used in experiments, we obtain the equation, connecting respective longevities τ and τ_0:

$$\int_0^{\tau_0} \frac{dt}{e^{-G\sigma_0(t)}} = \int_0^{\tau} \frac{dt}{e^{-G(\sigma_0(t) + \delta\sigma(t))}}.$$ (9)

Here $\sigma_0(t)$—linearly increasing main stress, $\delta\sigma(t)$—additional periodic stress of period T with amplitude $A_0/2$, having saw-tooth shape; $G = \gamma/RT_0$. Performing integration we find, that

$$\Delta = \Delta_0 \frac{1}{G\dot\sigma_0\Delta_0} \ln\left\{1 + \frac{1}{Y}(e^{\,G\dot\sigma_0\Delta_0} - 1)\right\},$$ (10)

where $\Delta = \tau - t_0$, $\Delta_0 = \tau_0 - t_0$, t_0—moment, when periodic loading has been turned on; $\dot\sigma_0$ and $\dot\sigma$—the rates of increasing of main and additional (periodic) loads, respectively;

$$Y = \frac{\dot\sigma_1 \, e^{-y_0}}{2(\dot\sigma_0 + \dot\sigma)}\left\{\frac{\cosh y}{\cosh y_0} + \frac{\sinh y}{\sinh y_0}\right\} + \frac{\dot\sigma_0 \, e^{y_0}}{2(\dot\sigma_0 - \dot\sigma)}\left\{\frac{\cosh y}{\cosh y_0} + \frac{\sinh y}{\sinh y_0}\right\},$$ (11)

$y = G\dot\sigma T_0/4 = GA_0/2$—the value proportional to the variation of additional periodic load during a quarter of its period, $y_0 = G\dot\sigma_0 T_0/4$—variation of main load during the same time interval.

Parameters, entering in (10), have the following values in conducted experiments: $\dot{\sigma}_0 = 1.2 \cdot 10^3$ Pa/s, $A_0 = 1.2 \cdot 10^5$ Pa, $\dot{\sigma} = 2.4 \cdot 10^6$ Pa/s, $\Delta_0 = \tau_0 - t_0 = 60$ s. The value γ was not measured in experiments directly. It is unfavorably known from literature data and strongly varies in experiments with rocks under compression and tension (STAVROGIN and PROTOSENYA, 1985; PETROV, 1993). Adopting the value $\gamma = 0.28$ kcal/mol/MPa, taken centrally over the band of known values, we will obtain $G = 4.7 \cdot 10^{-7}$ Pa^{-1}. Thus we have: $\dot{\sigma}_0 \ll \dot{\sigma}$, $y_0 = 1.4 \cdot 10^{-5} \ll 1$, $y = 0.03 \ll 1$, $G\dot{\sigma}_0\Delta_0 = 0.03 \ll 1$. The expressions (10) and (11) are simplified, taking into account the smallness of these parameters:

$$Y \approx \frac{y_0}{y}\{\sinh y \coth y_0 - \cosh y \tanh y_0\} \approx 1 + \frac{y^2}{6} - \frac{y_0^2}{y},$$

and

$$\Delta \approx \frac{\Delta_0}{Y}.$$

From here we derive the expression for a normal difference of times of instability appearance in experiments without vibration and with one

$$\frac{\Delta_0 - \Delta}{\Delta_0} = \frac{y^2}{6} - \frac{y_0^2}{y}. \tag{12}$$

The value of a vibration period only enters in the term $y_0 = G\dot{\sigma}_0 T/4$. In our case $y_0 = 1.4 \cdot 10^{-5}$, and $y = GA_0/2 = 0.03$, i.e., second term in (12) is considerably less than first. The influence of vibration frequency to a decrease of time interval to instability is insignificant, and its value is proportional to the square of vibration amplitude. It does not contradict the experimental results described above. It follows from (12) that vibration added by loading always influences in the direction of decrease of time interval to instability. The quantitative estimation is now impossible, owing to a lack of reliable information regarding the value of γ. It should be mentioned, however, that the absence of the clear empirical relation between T_i and the duration of vibration leaves some doubts about the role of durability in above described experiments.

Finally, the third mechanism of T_i reduction is not impossible in our case. The stick-slip can occur at a certain threshold stress level. In the experiments with vibration this level was reached earlier because of local maxima of the sinusoidal stress. When the half amplitude of vibration is about 7–8% of the critical stress, then it corresponds to the recorded reduction of T_i.

It is evident that the low-frequency vibration in the frequency range of 1–30 Hz did not cause the reciprocations of blocks along the contact, as happened when the stick-slip was initiated by high-frequency shock. In the latter case, the length of the elastic wave of initiating impulse is commensurable with the size of the plane of the contact, which in fact might be decisive.

The accomplished studies imply at least one geophysical application. The existence in the earth of vibrations of stresses with different frequencies, including those generating by earth tides, should be taken into account when practicing the model for forecasting the time and magnitude of recurring earthquakes.

Acknowledgements

The authors extend their sincere thanks to O. Babichev, V. Los', M. Grigor'ev, A. Patonin, and V. Terent'ev for their assistance in the experiments and V. Babicheva, E. Irisova, M. Preobrazhenskaya for their cooperation in data processing.

The work was supported by the Russian Foundation for Basic Research.

REFERENCES

BODIN, P., and GOMBERG, J. (1994), *Triggered Seismicity and Deformation between the Landers, California, and Little Skull Mountain, Nevada, Earthquakes*, Bull. Seismol. Soc. Am. *84*, 835–843.

DIETERICH, J. H. (1979), *Modeling of Rock Friction. 1. Experimental Results and Constitutive Equations*, J. Geophys. Res. *84*, 3162–2168.

PETROV, V. A. (1993), *Long-term Prediction of Earthquake Probability*, Modelling of the Seismic Process and Earthquake Precursors M., 48–52 (in Russian).

REGEL, V. P., SLUZKER, A. I., and TOMASHEVSKI, E. E. (1974), *Kinetic Basis of Solids Strength*, M., Nauka *560* (in Russian).

SHIMAZAKI, K., and NAKATA, T. (1980), *Time-predictable Recurrence Model for Large Earthquakes*, Geophys. Res. Lett. *7*, 279–282.

SOBOLEV, G. A., KOLTSOV, A. V., and ANDREEV, V. O. (1991), *Trigger Effect of Oscillations in Earthquake Models*, Dokl. Ac. Nauk USSR *319*, No. 2, 337–341 (in Russian).

SOBOLEV, G., SPETZLER, H., KOLTSOV, A., and CHELIDZE, T. (1993), *An Experimental Study of Triggered Stick-slip*, Pure and Appl. Geophys. *140*, 79–94.

STAVROGIN, A. N., and PROTOSENYA, A. G. (1985), *The Strength of Rocks and Stability of Mines*, M., Nedra *271* (in Russian).

(Received February 2, 1995, revised October 25, 1995, accepted December 11, 1995)

PAGEOPH, Vol. 147, No. 2 (1996)

0033–4553/96/020357–10$1.50 + 0.20/0

Decomposition of Seismic Moment Tensors for Underground Nuclear Explosions

Z. L. Wu[1] and Y. T. Chen[1]

Abstract —Generally the decomposition of a seismic moment tensor is not unique. However, to favorably view the characteristics of a certain seismic source, one must decompose a seismic moment tensor into parts according to assumptions about the properties of the seismic source. Different from natural earthquakes in which the shear dislocation component plays a predominant role in the source process, and the seismic moment tensor can be separated into an isotropic component, a double couple, and a compensated linear vector dipole (CLVD), underground nuclear explosions have three major components in their source process, i.e., the explosion, the tensional spalling, and the tectonic strain release associated with the explosion. In such a situation the conventional moment tensor decomposition for earthquakes is not convenient to estimate the yield of the explosion and to characterize the tectonic strain release. In this paper, an alternative decomposition scheme is proposed to deal with the moment tensor of underground nuclear explosions, which might benefit the approach to study the tectonic strain release induced by underground nuclear detonations.

Key words: Seismic moment tensor, underground nuclear explosion, tectonic strain release.

Introduction

Associated with underground nuclear explosions are two important secondary sources, i.e., the tensional spall and the tectonic strain release. Various studies have been carried out on the properties of these two secondary sources (e.g., TOKSÖZ and KEHRER, 1971, 1972; DAY et al., 1983; WALLACE et al., 1983, 1985; COHEE and LAY, 1988; PATTON, 1990; DAY and McLAUGHLIN, 1991; EKSTROEM and RICHARDS, 1994). In such studies the moment tensor approach is of special interest. Since the concept of seismic moment tensor was introduced (GILBERT, 1970, 1973) and extended (BACKUS and MULCAHY, 1976, 1977; BACKUS, 1977a,b), many studies have been conducted on the characterization of nuclear or chemical explosion sources in terms of seismic moment tensors (e.g., STUMP and JOHNSON, 1981, 1984, 1987; JOHNSON, 1988; PATTON, 1988; VASCO, 1989, 1990). As the first-order approximation of the description of a general seismic source, moment tensor is presumed to be one of the most useful tools in the approach to discriminate

[1] Institute of Geophysics, State Seismological Bureau, 100081 Beijing, China.

between earthquakes and underground nuclear explosions, to estimate the explosion yield, and to characterize the induced seismicities associated with the explosions.

In terms of seismic moment tensors, the characterization of the secondary sources falls into the decomposition of a general seismic moment tensor. In principle, such a decomposition is not unique (KNOPOFF and RANDALL, 1970). However, to observe the characteristics of a certain seismic source, one must separate the moment tensor into parts according to assumptions associated with the characteristics of the seismic source. Generally in dealing with an earthquake, one of the most commonly cited techniques is to find the eigenvalues and eigenvectors of the moment tensor (KNOPOFF and RANDALL, 1970; FITCH et al., 1980). In the principal coordinate system defined by the eigenvectors, the moment tensor is diagonalized and can be separated into three parts: an isotropic part, a double couple, and a compensated linear vector dipole (CLVD). This decomposition scheme plays an important role in the study of earthquake sources. However, as can be shown below, such a decomposition scheme may contain problems in dealing with underground nuclear explosions. In this paper we suggest an alternative scheme to deal with the moment tensors of underground explosions. Such a theoretical approach is related to two problems in practice: the first one is to accurately estimate the yields of explosions using a moment tensor inversion technique, and the second one is to retrieve the geometry of the tectonic strain release which occurs almost simultaneously with the explosion.

Moment Tensor Decomposition for Earthquakes

Often in earthquake seismology the seismic moment tensor is decomposed into an isotropic component, a double-couple, and a compensated linear vector dipole (CLVD):

$$\begin{bmatrix} M_1 & 0 & 0 \\ 0 & M_2 & 0 \\ 0 & 0 & M_3 \end{bmatrix} = EP \begin{bmatrix} 1 & 0 & 0 \\ 0 & 1 & 0 \\ 0 & 0 & 1 \end{bmatrix} + DC \begin{bmatrix} 1 & 0 & 0 \\ 0 & 0 & 0 \\ 0 & 0 & -1 \end{bmatrix} + CLVD \begin{bmatrix} -1 & 0 & 0 \\ 0 & 2 & 0 \\ 0 & 0 & -1 \end{bmatrix} \quad (1)$$

in which

$$EP = \tfrac{1}{3}(M_1 + M_2 + M_3)$$

$$DC = \tfrac{1}{2}(M_1 - M_3)$$

$$CLVD = \tfrac{1}{6}(2M_2 - M_3 - M_1).$$

In such a decomposition, the isotropic part is regarded as related to the volumetric change, and the double couple is related to the shear dislocation produced by the rupture of rocks. To some extent, such a decomposition scheme may be referred to as a kind of 'double-couple-oriented' scheme. For earthquakes in which most of the shear dislocation plays the predominant role, it is convenient to decompose the moment tensor into these three components. Despite that literature exists which deals with the non-double-couple components of earthquakes and their spatial distribution, and it is argued that there might be large volumetric changes associated with some earthquakes (e.g., YAO et al., 1994), what we are mainly concerned about is the shear dislocation, and either the volumetric change and the CLVD component plays the role of less importance. In fact in some CMT inversion algorithms, a criterion is used that non-double-couple components of the moment tensor minimize, as one of the criteria in the selection of the centroid depth. Also in most cases, in dealing with earthquakes we often treat the small secondary contribution of the isotropic component and the CLVD component as a result of the incompleteness of the structure model and the errors in the inversion (e.g., WU et al., 1994a). All these concepts originate from the fact that for most earthquakes, the non-double-couple components are found to be much smaller than the primary double-couple component.

Underground nuclear or chemical explosions, however, are different. It is well known that in underground explosions there are three major sources, i.e., the explosion, the tensional spall, and the tectonic strain release associated with the explosion (see, e.g., WU et al., 1994b). Within some frequency band, these three sources contribute altogether to the excitation of seismic waves. In such a situation, the description of a nuclear or chemical explosion, in terms of seismic moment tensors, plays a key role in the approach to characterize the source processes of the explosion and its secondary sources. And problems surface when we treat the moment tensors of explosions using the conventional decomposition scheme.

Decomposition of Seismic Moment Tensors for Underground Nuclear Explosions

In underground nuclear explosions there are three major sources: the explosion, the tensional spall, and the tectonic strain release associated with the explosion (e.g., WU et al., 1994b). In the nomenclature of the seismic moment tensor, the explosion itself can be represented as an isotropic moment tensor $EP\delta_{ij}$, in which EP represents the intensity of the explosion.

The second part of the moment tensor is the tectonic strain release caused by the explosion. It can be modeled, at least in the sense of the first-order approximation, as a double-couple.

Associated with underground nuclear explosions, there is another important source, i.e., the tensional spall caused by the upward propagated shock waves and the downward propagated free-surface reflections. In such a source there is a tensional crack, the moment tensor of which includes an isotropic part and a deviatoric part. This indicates that when we treat the explosion as a seismic source and decompose it into an isotropic part, a double couple, and a CLVD component, we actually have divided the tensional moment tensor into two parts, with one part given to the isotropic part, causing the uncertainty in the yield estimation, and another part to the deviatoric moment tensor, which contribute to both the double couple and the compensated linear vector dipole, causing the distortion of the geometry of the double couple.

Having noticed that the moment tensor consisting of an explosion and a double couple will have a zero CLVD component, in our approach, we attribute the compensated linear vector dipole of explosions to the contribution of spall, and propose an alternative decomposition scheme:

$$M_{ij} = EP\delta_{ij} + SP_{ij} + DC_{ij} \tag{2}$$

in which the first term on the right-hand side is the explosion itself, the second term is the spall, and the third term is the double couple. For the condition of underground nuclear explosions, the spall arises from the tensional dislocation of rocks under the tensional stress resulting from the upward propagated shock waves and the downward propagated free-surface reflections. Taking the free surface nearly horizontal, the spall term may be represented as (DAY and McLAUGHLIN, 1991)

$$SP_{ij} = \begin{bmatrix} \lambda DA & 0 & 0 \\ 0 & \lambda DA & 0 \\ 0 & 0 & (\lambda + 2\mu)DA \end{bmatrix} \tag{3}$$

in which λ and μ are the Lamé's constants, and D and A are the spall separation and the spall area, respectively. Taking the V_P/V_S ratio from other observations, the relative values of the matrix elements will be obtained, and the only variable to be determined is the amplitude. For example, taking $V_P/V_S = \sqrt{3}$, one has

$$SP_{ij} = SP \begin{bmatrix} 1 & 0 & 0 \\ 0 & 1 & 0 \\ 0 & 0 & 3 \end{bmatrix}. \tag{4}$$

The third term on the right-hand side of equation (2), as a double-couple source, obeys the constraints:

$$DC_{11} + DC_{22} + DC_{33} = 0 \tag{5}$$

$$\begin{vmatrix} DC_{11} & DC_{12} & DC_{13} \\ DC_{21} & DC_{22} & DC_{23} \\ DC_{31} & DC_{32} & DC_{33} \end{vmatrix} = 0. \tag{6}$$

The solution of the above equations is straightforward.

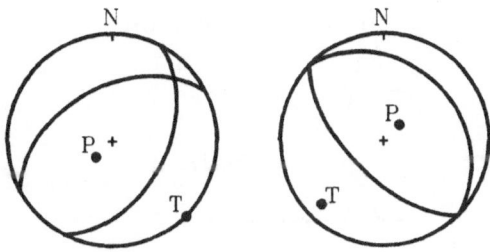

Figure 1
Double couple obtained using different decomposition schemes. See text for details.

Results of Digital Experiments

In dealing with underground nuclear explosions, we often hope to retrieve the time history and the geometry of the tectonic strain release as well as the time history of the spall and the explosion. However, if we still utilize the conventional decomposition scheme, resultingly we will obtain a volumetric change, a double couple, and a compensated linear vector dipole, in which the geometry of the best double couple will be different from that of the tectonic release, and the amplitude of the isotropic part will differ from that of the explosion.

Figure 1 provides a theoretical illustration in which we are going to demonstrate what happens when we handle an explosion source using the conventional decomposition scheme. In the example we are involved with a moment tensor of

$$M = \begin{bmatrix} 1.0 & 0.1 & 0.1 \\ 0.1 & 1.0 & 0.1 \\ 0.1 & 0.1 & 1.5 \end{bmatrix}.$$

It is quite similar to the moment tensor inversion results for chemical and/or nuclear explosions (e.g., STUMP and JOHNSON, 1987), in which clearly we can see the effects of the explosion and the spall, also there are perturbations with the moment tensor.

With the decomposition scheme of explosion, spall, and double couple, we have the decompositon result $EP = 0.50$, $SP = 0.40$, and $DC = 0.24$. On the other hand, if we separate the moment tensor into an explosion, a double couple, and a compensated linear vector dipole, we have $EP = 1.17$, $DC = 0.32$, and $CLVD = -0.06$, indicating how the explosion yield and the shear dislocation component are overestimated. In Figure 1 the left-hand side is the double couple in the conventional decomposition scheme of EP, DC and $CLVD$, while the right-hand side is the double couple in the decomposition scheme of EP, SP and DC. It may be seen that using different decomposition schemes, the geometries of the

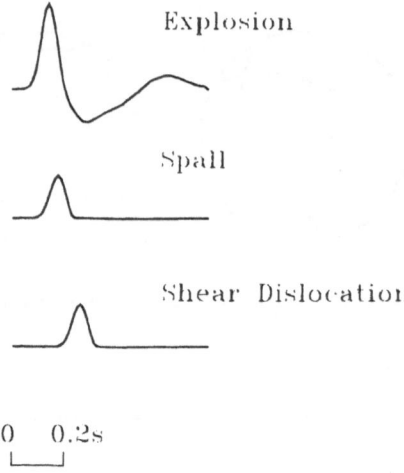

Figure 2
Source time function of the explosion, the spall and the tectonic strain release for the input source.

double couple are quite different. This fact may be of importance in the characterization of the tectonic strain release associated with underground nuclear explosions. Even more interesting is that in the geometry of seismic moment tensor
representation, even some invariants which are independent of the coordinate
systems, e.g., the isotropic component of the moment tensor (its trace), still contain
ambiguities which depend on the assumptions of the seismic sources.

It is believed that spall and tectonic strain release work in different frequency
bands and affect the seismic radiations within different frequency ranges. However,
sometimes it is possible that within some frequency range, the explosion, the spall,

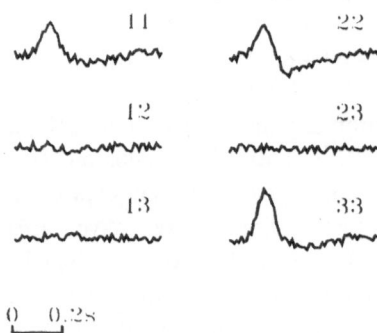

Figure 3
Synthetic time-dependent moment rate tensor for the experiment.

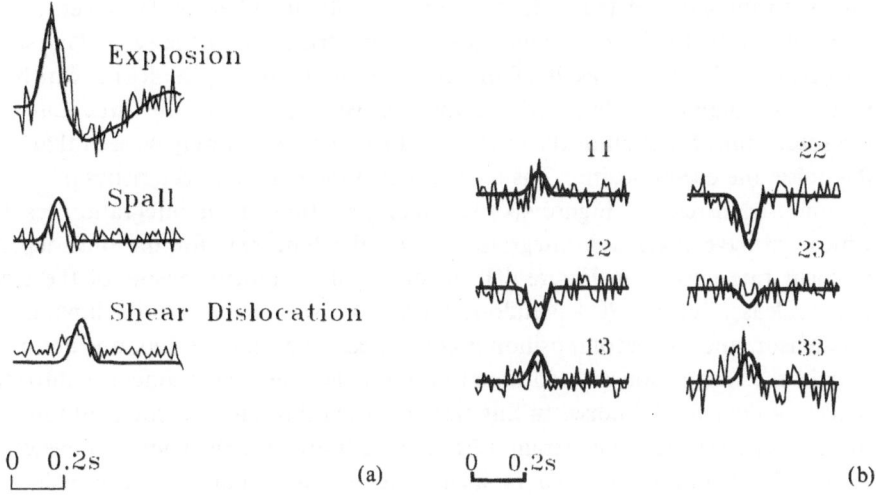

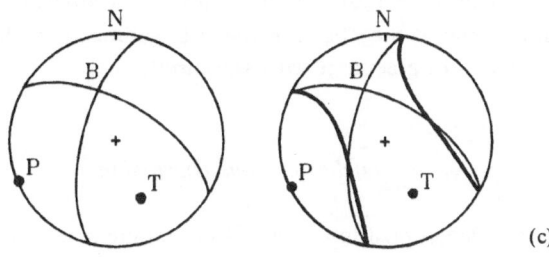

Figure 4

Result of decomposition. (a) Retrieved source time functions of the *EP*, *SP* and *DC* components (thin lines) and the input source time functions (thick lines). (b) Retrieved time-dependent moment rate tensor of the *DC* component (thin lines) and the input moment rate tensor of the *DC* component (thick lines). (c) Geometry of the input tectonic strain release (left) and the retrieved result, see text for details.

and the shear dislocation may exist simultaneously and contribute altogether to the excitation of seismic waves. To demonstrate our decomposition scheme for this complex situation we carried out numerical experiments with the synthetic data. The synthetic data are taken similar to the inversion results in real cases. Figure 2 shows the source time functions of the explosion, the spall and the tectonic strain release for the input source. In this example the double couple describing the tectonic strain release is taken as strike 120°, dip 60°, and rake 30°. The amplitude of the three sources are taken as $EP = 1.0$, $SP = 0.5$, and $DC = 0.5$. Also, to reflect the incompleteness of the structure model and the errors in the inversion, the combined moment tensor is purposely contaminated by 20% pseudo-random noise. The synthetic time-dependent moment rate tensor is displayed in Figure 3, which is quite similar to the moment tensor inversion results obtained before. However, from the moment tensor it is hard to differentiate the contribution of the secondary sources.

Figure 4 shows the result of the decomposition. In Figure 4(a) it can be seen that the decomposition has retrieved the relative amplitude of the three sources and the source time function of the explosion. In Figure 4(b) it may be seen that despite the noise, the decomposition has reconstructed the main characteristics of the input tectonic strain release. Figure 4(c) is a representation of the integration results, in which we have taken the integration along the time axis for the time-dependent moment rate tensors in Figure 4(b) to obtain the moment tensors of the tectonic strain release. Because 20% pseudo-random noise was included in the input moment rate tensor, and the decomposition is conducted for each time step, it is not peculiar that in the integration an additional non-double-couple component is introduced, reflecting the effect of noise. In this case, for each time step the moment rate tensor of the retrieved tectonic strain release is a pure double couple, however, the temporal integration of such moment rate tensor is not a pure double couple because the orientation of the retrieved shear dislocation fluctuates with time due to the random noise. In Figure 4(c), for the retrieved result on the right-hand side we also show the double couple (in thin lines) which is nearest to the deviatoric part of the moment tensor (the best double couple). Comparing the decomposition result and the input one, it may be seen that the geometry of the retrieved tectonic strain release is in good agreement with the input.

Conclusions and Discussion

In principle, the decomposition of a seismic moment tensor is not unique. As a result, different decomposition schemes based on different assumptions regarding the source process may lead to different conclusions. Different from earthquakes in which the shear dislocation component plays a predominant role in the source process, and the seismic moment tensor can be separated into an isotropic component, a double couple, and a compensated linear vector dipole, underground nuclear explosions have three major components in their source process, i.e., the explosion, the tensional spall, and the tectonic strain release associated with the explosion. In such a condition, the conventional moment tensor decomposition for earthquakes is not appropriate to estimate the yield of the explosion and to characterize the tectonic strain release. In this paper another decomposition scheme is proposed to deal with the moment tensor of underground nuclear explosions. In such a decomposition scheme the moment tensor is separated into an explosion, a double couple, and a spall. We have shown that by using different decomposition schemes, the amplitude of the explosion and the geometry of the double couple may be very different. Our results indicate that to some extent, the problem of moment tensor decomposition is a physical problem rather than a mathematical problem.

In this paper it is assumed that the point source approximation holds valid and the moment tensor of an explosion has been well inverted from seismic wave form data. In this case the spatial difference of the explosion, the spall and the tectonic strain release are compensated into their temporal difference, and the point source acts as the centroid. In practice, as has been demonstrated in previous works, to obtain the reliable moment tensors of explosions is not easy, and will continue to be an important topic in nuclear explosion seismology. Our paper has set up another question, i.e., even if the moment tensor has been well reconstructed, to obtain an appropriate description pertaining to the secondary sources, one must consider the decomposition scheme to be chosen in the analysis.

Acknowledgements

Thanks are due to Dr. Peter Knoll for his help and to the anonymous reviewers for their constructive suggestions.

REFERENCES

BACKUS, G. (1977a), *Interpreting the Seismic Glut Moments of Total Degree Two or Less*, Geophys. J. R. Astron. Soc. *51*, 1–25.

BACKUS, G. (1977b), *Seismic Sources with Observable Glut Moments of Spatial Degree Two*, Geophys. J. R. Astron. Soc. *51*, 27–45.

BACKUS, G., and MULCAHY, M. (1976), *Moment Tensors and Other Phenomenological Descriptions of Seismic Sources, I, Continuous Displacements*, Geophys. J. R. Astron. Soc. *46*, 341–361.

BACKUS, G., and MULCAHY, M. (1977), *Moment Tensors and Other Phenomenological Descriptions of Seismic Sources, II, Discontinuous Displacements*, Geophys. J. R. Astron. Soc. *47*, 301–329.

COHEE, B. P., and LAY, T. (1988), *Modeling Teleseismic SV Waves from Underground Explosions with Tectonic Release: Results for Southern Novaya Zemlya*, Bull. Seismol. Soc. Am. *78*, 1158–1178.

DAY, S. M., RIMER, N., and CHERRY, J. T. (1983), *Surface Waves from Underground Explosions with Spall: Analysis of Elastic and Nonlinear Source Models*, Bull. Seismol. Soc. Am. *73*, 247–264.

DAY, S. M., and MCLAUGHLIN, K. L. (1991), *Seismic Source Representations for Spall*, Bull. Seismol. Soc. Am. *81*, 191–201.

EKSTROEM, G., and RICHARDS, P. G. (1994), *Empirical Measurements of Tectonic Moment Release in Nuclear Explosions from Teleseismic Surface Waves and Body Waves*, Geophys. J. Intl. *117*, 120–140.

FITCH, T. J., MCCOWAN, D. W., and SHIELDS, M. W. (1980), *Estimation of the Seismic Moment Tensor from Teleseismic Body Wave Data with Applications to Intraplate and Mantle Earthquakes*, J. Geophys. Res. *85*, 3817–3828.

GILBERT, F. (1970), *Excitation of the Normal Modes of the Earth by Earthquake Sources*, Geophys. J. R. Astron. Soc. *22*, 223–226.

GILBERT, F. (1973), *Derivation of Source Parameters from Low-frequency Spectra*, Phil. Trans. R. Soc. *A274*, 369–371.

JOHNSON, L. R. (1988), *Source Characteristics of Two Underground Nuclear Explosions*, Geophys. J. *95*, 15–30.

KNOPOFF, L., and RANDALL, M. J. (1970), *The Compensated Linear Vector Dipole: A Possible Mechanism for Deep Earthquakes*, J. Geophys. Res. *75*, 4957–4963.

PATTON, H. J. (1988), *Source Models of the Harzer Explosion from Regional Observations of Fundamental-mode and Higher-mode Surface Waves*, Bull. Seismol. Soc. Am. *78*, 1133–1157.

PATTON, H. J. (1990), *Characterization of Spall from Observed Strong Ground Motions on Pahute Mesa,* Bull. Seismol. Soc. Am. *80,* 1326–1345.

STUMP, B. W., and JOHNSON, L. R., *The effect of Green's functions on the determination of source mechanisms by the linear inversion of seismograms.* In *Identification of Seismic Sources: Earthquake or Underground Explosion* (Husebye, E. S., and Mykkeltveit, S., eds.), (D. Reidel Publishing Company 1981) pp. 255–268.

STUMP, B. W., and JOHNSON, L. R. (1984), *Near-field Source Characterization of Contained Nuclear Explosions in Tuff,* Bull. Seismol. Soc. Am. `74,` 1489–1502.

STUMP, B. W., and JOHNSON, L. R. (1987), *Mathematical Representation and Physical Interpretation of a Contained Chemical Explosion in Alluvium,* Bull. Seismol. Soc. Am. 77, 1312–1325.

TOKZÖS, M. N., and KEHRER, H. H. (1971), *Underground Nuclear Explosions: Tectonic Utility and Dangers,* Science *173* (3993), 230–233.

TOKZÖS, M. N., and KEHRER, H. H. (1972), *Tectonic Strain Release by Underground Nuclear Explosion and its Effect on Seismic Discrimination,* Geophys. J. R. Astron. Soc. *31,* 141–161.

VASCO, D. W. (1989), *Deriving Source-time Functions Using Principle Component Analysis,* Bull. Seismol. Soc. Am. *79,* 711–730.

VASCO, D. W. (1990), *Moment Tensor Invariants: Searching for Non-double-couple Earthquakes,* Bull. Seismol. Soc. Am, *80,* 354–871.

WALLACE, T. C., HELMBERGER, D. V., and ENGEN, G. R. (1983), *Evidence of Tectonic Release from Underground Nuclear Explosions in Long-period P Waves,* Bull. Seismol. Soc. Am. *73,* 593–613.

WALLACE, T. C., HELMBERGER, D. V., and ENGEN, G. R. (1985), *Evidence of Tectonic Release from Underground Nuclear Explosions in Long-period S Waves,* Bull. Seismol. Soc. Am. *75,* 157–174.

WU, Z. L., CHEN, Y. T., NI, J. C., WANG, P. D., and WANG, M. (1994a), *Moment Tensor Inversion of Near-source Broadband Data,* Acta Seismologica Sinica (English edition) *7,* 187–199.

WU, Z. L., CHEN, Y. T., and MU, Q. D. (1994b), *Nuclear Explosion Seismology: An Outline,* Seismological Press, 19–46 (in Chinese).

YAO, Z. X., ZHENG, T. Y., and WEN, L. X. (1994), *Moment Tensor Inversion Method for Determining the Earthquake Process by Use of P-waveform Data,* Acta Geophysica Sinica *37,* 36–44 (in Chinese with English abstract).

(Received January 25, 1995, revised October 10, 1995, accepted December 11, 1995)

PAGEOPH, Vol. 147, No. 2 (1996)

0033–4553/96/020367–09$1.50 + 0.20/0

Invariant Kinetic Approach to the Description of a Rock Fracture Process and Induced Seismic Events

Valery A. Anikolenko[1] and Vladimir A. Mansurov[2]

Abstract —Powerful seismic events, such as earthquakes and rockbursts, are caused by the accumulation of energy in rocks and loss of rock mass stability. Usually methods of their forecasting are based on the registration of anomalous behavior of geophysical fields. However an efficiency of this approach is low. The present paper proposes a kinetic approach to the description of rock fracture process, which can be used for the forecasting of seismic events and an investigation of structure and energy distributions in rock. 3-D and 1-D kinetic equations describing a process of cluster formation in rock were obtained. The equations are invariant to deformation conditions and to the scale level of events. They showed a good agreement with the results of field observations and laboratory experiments. It was also shown that these equations well describe the processes of earthquake, rockburst and rock sample failure preparation. Catalogues of rockbursts in mines were analyzed with the use of the kinetic equations to find out evidence of induced seismic events. The proposed approach makes it possible to reveal trends in rock behavior and thus predict the rock failure at different scale levels.

Key words: Induced seismicity, kinetics, rock fracture, rockburst, earthquake.

1. Introduction

Practical needs of mining industries and the burying of radioactive and chemical waste require long-term monitoring of stressed state in a rock massif to prevent accidents which may be caused by rockbursts, earthquakes, landslides, etc. In many cases, dense instrumental network capable of registering excessive local stresses or deformations in a rock massif could be replaced or supported by relatively simple registration systems which can monitor trends in stressed media for a sufficiently long time. However, use of such systems demands a better understanding of the rheological properties and physical mechanism of the rock failure process.

The problem can be reduced to defining quantitative physical criteria of stability loss by a rock massif, in other words, to finding earthquake or rockburst predecessors. Usually these predecessors are understood as anomalies of a certain measured

---

[1] Institute of Experimental Geophysics, United Institute of Physics of the Earth, Russian Academy of Science, 51 Nikoloyamskaya St., Moscow 109004, Russia.
[2] Institute of Physics and Mechanics of Rock, Academy of Sciences of Kirgizstan, 98 Mederova St., Bishkek 720035, Republic of Kirgizstan.

physical field short time before the seismic event has happened. However, a large number of predecessors which do not often correlate with one another, and are local in space and time, makes this approach inefficient for the forecasting of a strong seismic event, because it does not take into consideration the history of its preparation. It is clear that the final stage of the event can develop in different ways due to its highly stochastic nature. The aim of this paper is to apply the theory and methods of physical kinetics to a description of rock mass failure preparation and the forecasting of seismic events. Kinetic approach to the description of a fracture process was first formulated by ZHURKOV (1957) for polymers and was later developed by REGEL et al. (1974), PETROV (1981) and others for rocks. This approach was also successfully used by MIKHAILOV (1971), ANIKOLENKO and MIKHAILOV (1976) in solid state physics and chemistry.

2. Formulation of the Problem and the Physical Model

Dynamic effects in rocks, such as earthquakes or rockbursts come as a result of: (i) the development of the deformation and failure process, which proceed in time, (ii) the localization of these processes in space; and (iii) the concentration of elastic energy in some areas, where relaxation is quite difficult. The loss of stability at such a failure center is accompanied by the rapid release of accumulated elastic energy, and it forms elastic impulses or waves which can possess destructive power. These general conditions were formulated in the works of AVERSHIN (1955), BLAKE (1972), COOK (1965), PETUKHOV (1972) and others.

As it is known, the crucial role in a fracture process belongs to microstresses, which exist *a priori* in all kinds of rocks due to the microheterogeneity and microanisotropy of geophysical media. These energy concentrators cause cracks which grow in number and size as the deformation develops. In this case rock mass can be interpreted as a statistical ensemble. During fracture process the ensemble overcomes a set of potential barriers of different energy, or width, which are due to interatomic interactions. A kinetic equation describing accumulation of defects with time can be written in the work by ANIKOLENKO and MIKHAILOV (1976):

$$N(t) = \int f(r)\{1 - \exp[-k(r)t]\}\, dr. \tag{1}$$

According to CHELIDZE (1987) the rate constant of the joining of defects i and j into the defect of a higher scale level has the following form:

$$k(r) = k_0 \exp(-2r/a) \tag{2}$$

where r is the distance between these defects, and k_0 and a are constants. Normally the value of a is of the order of a minimal cluster size for a given rock type. The distribution $f(r)$ can be written in the following form (see ANIKOLENKO and MIKHAILOV, 1976; and ANIKOLENKO, 1992):

$$f(r) = 4\pi r^2/(4/3\pi r_0^3) \exp(-r^3/r_0^3) \tag{3}$$

where $r_0 = (3C/4\pi)^{-1/3}$ is a mean distance between the defects at the beginning of observations; C is the initial defect concentration. The distribution (3) shows a probability for the separate defect i to find a neighbor defect j at the radius r within the spherical layer dr thick. Then, substituting (2) and (3) into (1) we shall obtain:

$$\ln N(t)/\ln N(t_0) = (a/2r_0)^3[\ln^3(k_0 t) - \ln^3(k_0 t_0)] \tag{4}$$

where t_0 is the beginning time of the observations.

In the case of a uniform distribution

$$f(r) = 1/(r_{max} - r_{min}) \tag{5}$$

instead of (4) we shall obtain the following equation:

$$N(t)/N(t_0) = [a/2(r_{max} - r_{min})] \ln(t/t_0) \quad \text{for} \quad t \gg t_0, \tag{6}$$

which describes a change of the defect concentration as a linear function of the logarithm of the elapsed time. This is a case of in pairs interaction of defects, when a distance between two joining defects is much shorter than a distance to another pair of interacting defects.

The equations (4) and (6) were obtained with an assumption of the Markov process of a crack formation. This process is a chain of subsequent acts of the joining of separate defects into a defect of a higher scale level. In case of the joining of long cracks, every separate crack can be considered as two-point defects localized at the edges of the crack. In the Markov process features of a defect depend only on features of the parent defects and are independent of the parameters of other predecessors.

Solving the inverse problem for the equations (4) and (6) we can obtain the distribution $f(r)$, as well as a value of the parameter a, which characterize a rate of mechano-chemical reaction of rock fracture. This will be a subject of a separate paper. Here we shall restrict our consideration to the analysis of kinetic dependencies which can also provide useful information regarding the rock fracture process.

3. Comparison of the Experimental Results

With this purpose a series of rock failure laboratory experiments were carried out. Cylindrical rock samples, 60 mm in height and 30 mm in diameter, were uniaxially loaded by means of a hydraulic testing machine and tested until the sample failed. Acoustic emission signals, generated by forming defects, were registered by wide-band piezoreceivers and than analyzed. The details of the used experimental technique have been reported by MANSUROV (1994).

Accordingly let us consider the results of these experiments. Figure 1a shows a typical differential spectrum of acoustic emission signals corresponding to the

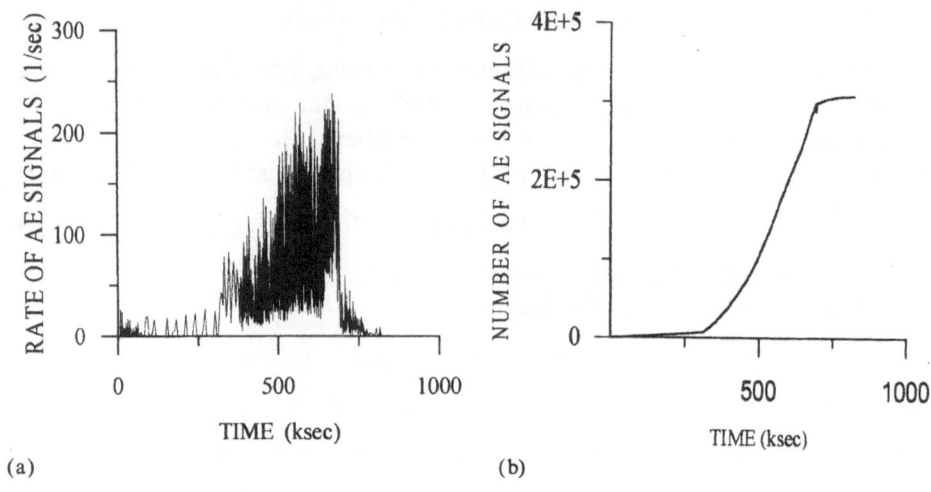

Figure 1
Failure of a sandstone sample using a testing machine: (a)—rate of acoustic emission (AE) signals vs.
time; (b)—cumulative number of AE signals vs. time.

failure of a sandstone sample under a uniaxial load. The load was applied during a relatively long time until the sample failed because of the formation of a major crack, that has split it into two main parts. Consequently, Figure 1b presents an integral spectrum of Figure 1a that shows a dependence of a cumulative number of acoustic emission signals versus elapsed time t. The experiments have also shown that differential spectra obtained for the samples at the same experimental conditions, differ essentially from one another even though they were made from the same block of rock. The obtained cumulative curves have a smooth shape. As was shown, their shape depends more on the kind of rock from which they were made, and less on the conditions of the loading in the experiment. All of these curves have an S-wise shape, and are reaching a plateau at the moment of the total failure of a sample.

Figure 2 illustrates the same curve plotted in the coordinates $N(t)$ versus $lg\ t$ of the equation (6). As is easy to see, more than 85 percent of the experimental points adhere to the linear dependence given by equation (6). The linear fragment of the kinetic curve is assumed to be due to in pair interactions of defects under the load and, correspondingly, to the generation of cracks of higher scale levels. Therefore, the starting nonlinear fragment of the kinetic curve corresponds to the stage of formation of primary (noninteracting) clusters, and the final (also nonlinear) segment corresponds to the formation of a major prefailure crack. A visible slowdown of the fracture process at the final stage of the experiment could be explained as a slowdown of the process of interpair interactions, when the major crack stops interactions of defects through its surface.

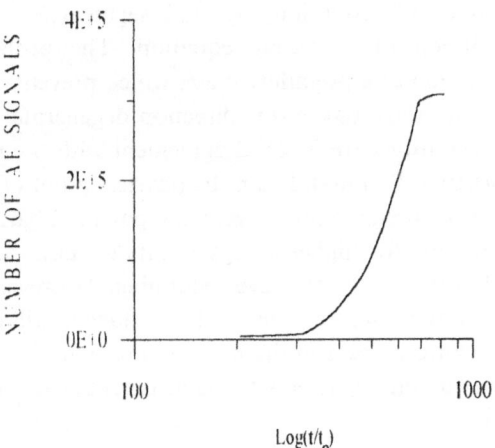

Figure 2
Figure 1b plotted as a function of logarithm of time.

It can be seen more illustratively from Figure 3, where a cumulative kinetic curve of a granite sample failure was plotted in the coordinates of the three-dimensional equation (4). Linear segments 1 and 2 can be identified, according to KUKSENKO (1984), as follows: the segment 1 is a stage of accumulation of initial defects, and the segment 2 is a stage of their by-pair interaction. The segment 3 of the curve evidently corresponds to the stage of formation of the major prefailure crack. As was established, segment 2 was well described by the equation (4), which

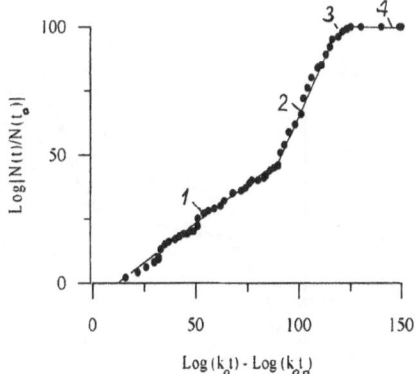

Figure 3
Kinetic dependence of failure of a granite sample using a testing machine. The kinetic curve is plotted in the coordinates of equation (4). Here $k_0 = 1$ s^{-1}, and t is measured in seconds. 1—generation of initial defects; 2—generation of higher level defects (cracks) due to their joining; 3—growth of the main prefailure crack; 4—failure of the sample.

is the 3-D solution of the equation (1), and segment 3 was described by the degenerate 2-D solution of the same equation. The probable reason of this degeneracy is a formation of a prefailure crack which prevents the growth of cracks through its surface and thus makes this direction degenerate.

The above considerations are in good agreement with the mechanical theory of rock fracture by SADOVSKY (1984). From the physical point of view this means that with time the statistical ensemble must overcome potential barriers of larger widths. The average waiting time for higher energy events increases and the process (Fig. 1b) seems to be slowing down, and even stopping. However, as it follows from equations (4) and (6), the fracture process at this stage is still in progress. A visible slowdown of the fracture process in the linear coordinates is due to the increase of waiting time for the joining of defects through potential barriers of larger widths.

4. Discussion

Finally, let us discuss the results of the experiments in which samples were tested together with iron parts of a similar cylindrical form. This allowed us to increase the stiffness of the testing machine, and thus to smooth out the dynamics of the sample failure during the final stage. Figure 4 shows a typical kinetic dependence for experiments of this kind. The curve consists of two segments, each is similar to the curve in Figure 1b. Each segment manifests an active growth of acoustic emission signals initially, then a slowdown is observed. The experiments demonstrated that the kinetic curve was self-reproducing every time a major prefailure crack, which has split the sample into separate fragments, was generated. Thus,

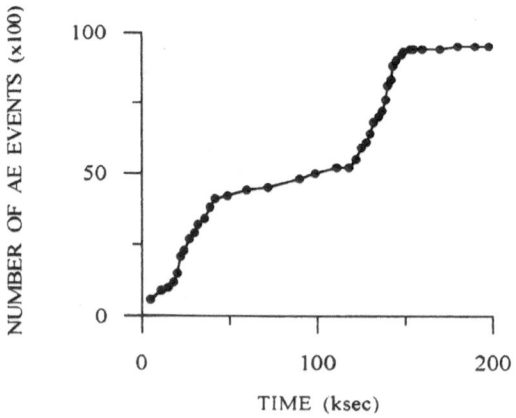

Figure 4
Dependence of cumulative number of AE signals vs. time (a two-stage failure of a marble sample).

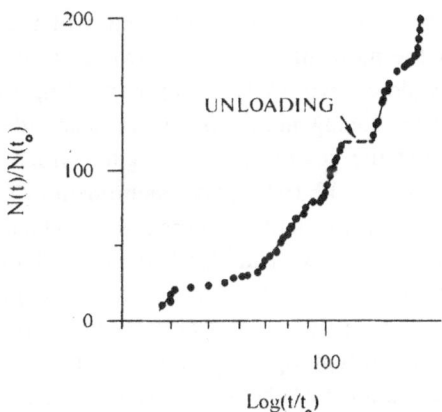

Figure 5
Cumulative number of AE signals vs. time (multi-stage failure of the marble sample).

every new stage of the reproduction of the kinetic curve corresponds to the failure of separate split fragments of the sample.

Figure 5 shows a semilog plot of the cumulative kinetic curve for the experiment, where a marble sample was tested as follows: (i) it was loaded with a testing machine for a relatively long time, then (ii) completely unloaded, (iii) loaded again until its failure. As is seen from Figure 5, the kinetic curve has five linear segments, each of which is following the law of the equation (6) as before, following the unloading of the sample. Simultaneous stress-strain measurements show a similar semilog dependence of stress on time. Evidently, this experiment proves the self-reproducing nature of the fracture process. The cumulative kinetic curve was reproduced in an invariant way several times during the test. Separate fragments of the sample, which were split during the first stage of the test, were split repeatedly into smaller fragments. In other words, every new linear fragment of this kinetic curve corresponds to the fracture process taking place at a new lower scale level. It is clear that after the unloading of the sample the kinetic curve starts from the same point and at the same parameters which have existed before the unloading of the sample.

The laboratory experiments which were considered above, evidence that the kinetic approach and the physical model can be used for the description of the seismic process which develops in rocks. Of course conditions of the laboratory experiments differ from natural conditions of the rock massif, which are more complicated. However, general conclusions made on the basis of laboratory experiments are also valid for natural conditions. The laboratory experiments simulate efficiently processes of accumulation and relaxation of elastic energy in rock mass under tectonic loads in a natural condition. The kinetic approach can be also applied to the separation of induced seismic phenomena from a regular seismicity:

cumulative kinetic curves plotted in the coordinates of the equations (4) and (6) have a sharp bend at the moment the induced event occurred. As a result of this, a seismic process may slow down or speed up depending on ambient conditions.

Evidently, a self-reproducing nature of defect accumulation and fracture processes is responsible for the so-called seismic gap, which often occurs before a powerful earthquake, or rockburst. This phenomenon results in a significant slowdown of the integral seismic activity in the area in which a powerful earthquake will then arise. A similar slowdown can also be seen in Figure 1b, where the kinetic curve at the prefailure stage bends and then comes to the plateau. It is clear that a seismic gap does not correspond to a real stop in seismic activity as it seems, because it is not observed in semilog coordinates. Moreover, the probability of high energy events increases as most of the low energy events have already taken place. From the point of view of earthquake forecasting, a seismic gap appears to be an efficient predecessor, especially in the case of a rupture crack formation, after which these events are followed by a full or partial energy release and a self-reproduction of the kinetic dependence similar to the one in Figure 5. However, in case of the energy release through the tectonic faulting, a seismic gap may not be observed at all, because such events have occurred only within the rising segment of the kinetic curve, ref. (Fig. 1b), i.e., before it has reached the plateau.

5. Conclusion

The results of laboratory and field experiments have demonstrated a good agreement with the theoretical physical model and the obtained kinetic equations. The proposed kinetic model can be used for the long-term forecasting of the time of occurrence of seismic events of any energy and scale level. Induced seismic phenomena can be separated from regular seismicity. The approach also provides a physical explanation of the well-known seismic phenomena, such as a seismic gap and the earthquake reoccurrence law. The kinetic model can be used as a base of a software for the earthquake and rockburst monitoring system.

REFERENCES

AVERSHIN, S. G., *Rockbursts* (Ugletechizdat, Moscow 1955) 235 pp. (in Russian).
ANIKOLENKO, V. A., *Kinetics of rock failure*. In *Physics and Mechanics of Rock Failure* (Nauka, Moscow 1992) (in Russian).
ANIKOLENKO, V. A. (1992), *Kinetic Study of Rock Fracture in Laboratory Press Experiments*, Acta Montana, A *3* (89), 63–72.
ANIKOLENKO, V. A., and MIKHAILOV, A. I. (1976), *Electron Tunneling in Glassy Organic Matrices*, Doklady AN SSSR *230*, 102–105 (in Russian).
BLAKE, W. (1972), *Rockbursts Mechanics*, Colo. Sch. Mines Q. *67*, 1–64.
CHELIDZE, T. L., *Methods of Percolation Theory in Mechanics of Geomaterials* (Nauka, Moscow 1987) 254 pp. (in Russian).

COOK, N. G. W. (1965), *A Note on Rockbursts Considered as a Problem of Stability*, J. S. Afr. Inst. Min. & Metall. *8* (65), 437–446.

KUKSENKO, V. S. (1984), *Kinetic Aspects of Rock Failure. Prediction of Earthquakes*, Dushanbe, USSR *4*, 8–12 (in Russian).

MANSUROV, V. A., *Laboratory experiments: their role in the problem of rock burst prediction.* In *Comprehensive Rock Engineering* (Pergamon Press 1994) 3, 745–771.

MIKHAILOV, A. I. (1971), *Decay of Defects in Irradiated Solids*, Doklady AN SSSR *197*, 136–140 (in Russian).

PETROV, V. A. (1981), *Principles of Kinetic Theory of Microfailure Prediction*, Fizika Tverdogo Tela *23*, 33–72 (in Russian).

REGEL, V. R., SLUTSKER, A. I., and TOMASHEVSKI, E. E. (1974), *Kinetic Nature of Strength in Solids*, Uspekhi Fizicheskikh Nauk *106*, 193–228 (in Russian).

SADOVSKI, M. A. (1984), *On Models of Geophysical Media and Seismic Process. Prediction of Earthquakes*, Dushanbe, USSR *4*, 268–273 (in Russian).

ZHURKOV, S. N. (1957), *On Problems of Strength in Solids*, Vestnik AN SSSR *11*, 78–82 (in Russian).

(Received January 18, 1995, revised September 30, 1995, accepted December 11, 1995)

PAGEOPH, Vol. 147, No. 2 (1996)

0033–4553/96/020377–11$1.50 + 0.20/0

The Character and Extent of Seismic Deformation in the Focal Zone of Gazli Earthquakes of 1976 and 1984, $M > 7.0$

L. M. PLOTNIKOVA,[1] B. S. NURTAEV,[2] J. R. GRASSO,[3] L. M. MATASOVA,[1] and R. BOSSU[3]

Abstract—With objective of investigating the peculiarities of seismic process development and seismotectonic deformation character in the focal area of the Gazli earthquakes of 1976 ($7.0 < M < 7.3$) and 1984 ($M = 7.2$), a local seismic network was installed. For the field observation period (May to June, 1991) more than 400 events with magnitudes $-0.2 < M < 4.5$ were recorded by at least 6 stations.

Isometric presentation of earthquake hypocenters distribution allows us to define the depth and dipping planes orientation of seismoactive faults of the region.

The focal mechanisms of 35 earthquakes for the period 1979–1988, $M > 2.8$, connected to a gas extraction regime period, and 75 events $1 < M < 4.3$ for the 1991 period (gas storage regime) are used to analyze the dynamics of seismotectonic deformation processes (SDP) in this region. It has been ascertained, that the earth's crust in the Gazli region is subject to complicated deformation processes, particularly below 4 km depth. The predominant kind of deformation is compression. Vertical velocities of deformation show uplift of most of the region during the period of field work. The maximum velocity of vertical deformations for the Gazli structure is $V = 0.41$ mm/year. The comparison of the vertical velocities' displacements due to seismic flow with recent tectonic movements of the earth's crust has revealed their direct relation and high percentage of seismic flow contribution to the tectonic movement. The results obtained testify that the active seismic processes in the Gazli region are connected not as much as the residual stress release in the focal zone of the earthquakes 05. 1976 and 1984, $M > 7.0$ but rather with the influence of the gas reservoir exploitational regime on the rocks with different rheologic properties.

Key words: Induced seismicity, gas field, seismic deformation.

Introduction

Understanding the mechanism of seismic imbalance in the regions of active gas and oil exploration (YERKES and CASTLE, 1976; GRASSO, 1990; PLOTNIKOVA *et al.*, 1990) is connected with the investigation of the following: the characteristics of the deep seismic faults, the dynamics of tectonic processes, the peculiarities of the

[1] Institute of Geology and Geophysics, Tashkent 700041, Uzbekistan.
[2] Institute of Seismology, Academy of Sciences of Uzbekistan.
[3] LGIT, IRIGM, Université Joseph Fourier, Grenoble, France.

seismotectonic deformations (STD) and the character of the changes of the deformation's components direction under the influence of different factors, including the gas reservoir exploitation regime. This allows us to work out a generalized statistical and geomechanical model of the phenomenon and to solve the problem of optimization of the gas reservoir exploitation regime purposeful to mitigating the damage from possible severe earthquakes.

Observations

Seismic monitoring was carried out in 1991 in the region of the Gazli gas field within the focal zone of 1976 and 1984 earthquakes, by means of 17 stations: 7 analog (MEQ 600) and 10 digital (Lytoscope) stations (PLOTNIKOVA *et al.*, 1994). It provided reliable statistical data which enabled us to establish morphological

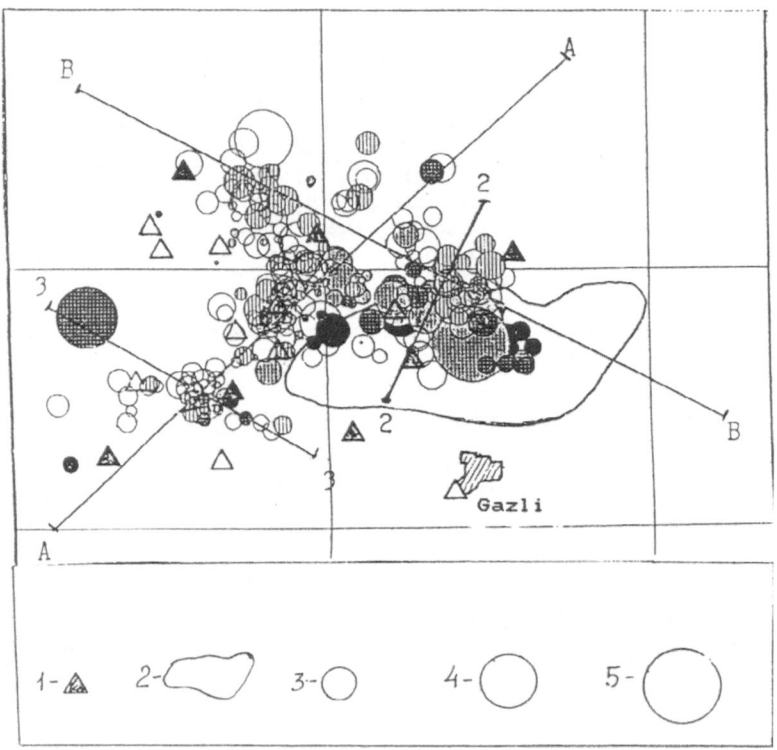

Figure 1
Seismicity map of Gazli region (May–June, 1991). 1—seismic stations; 2—contours of the gas field; classification of earthquakes by intensity: 3—$M = 2$; 4—$M = 3$; 5—$M > 4$.

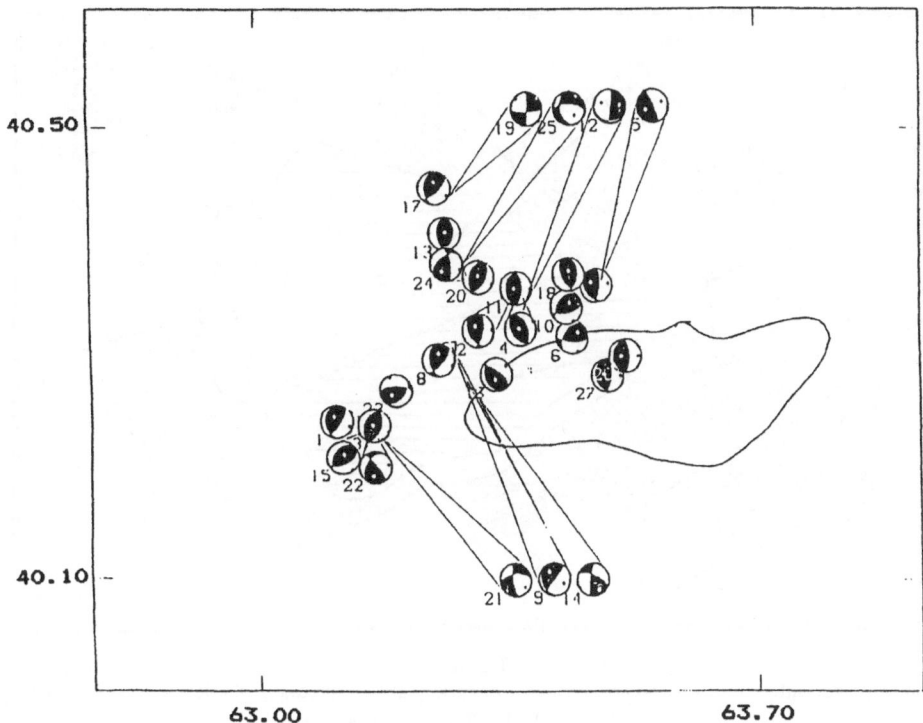

Figure 2
Focal mechanisms of earthquakes of Gazli region 2.5 < *M* < 4.5 (May–June, 1991).

parameters of tectonic faults and to analyze the seismotectonic deformations of the earth's crust in the region.

During the period of field work (May to June, 1991) 650 earthquakes with an intensity $-0.2 < M < 4.3$ were recorded. The accuracy of arrival time at the station is $=0.05$ s, and that of kinematic parameters of hypocenters is $=0.03$ s (Fig. 1).

The focal mechanisms of 35 earthquakes for the period from 1979 to 1988 $M > 2.8$ (gas extraction regime) and 75 earthquakes $1 < M < 4.3$ for the period of 1991 (gas storage regime) were used to analyze the influence of the gas field operation regime upon the distribution of the stress near the source (Fig. 2). From 1988 the gas reservoir has been used as the regional industrial gas storage; every year for 6 months (April–October) gas has been injected, and for 6 months (November–March) it has been extracted.

The parameters of tectonic faults (Fig. 3) in the Gazli gas field area were defined by means of isometric presentation of the observed events (PLOTNIKOVA *et al.*, 1991).

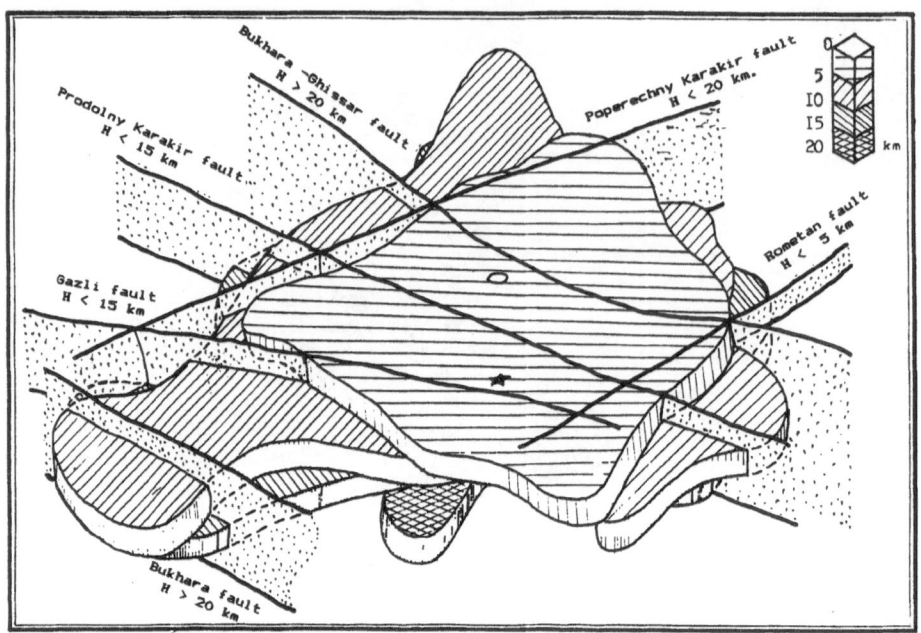

Figure 3
Isometric presentation of vertical cross sections of earthquakes' sources distribution $-0.2 < M < 4.3$ in
Gazli region (depth interval 0–20 km).

Characteristics of Seismotectonic Deformations

The estimation of seismotectonic deformations can be reduced to the problem of
calculating the seismic flow of the rock masses, based on integrating deformations
and deformation velocities in the sources within the temporal and spatial domain
and the determination of tensors of deformation components by one of the main
axes of the average mechanism (RIZNICHENKO, 1985; NIKITIN and YUNGA, 1977).
In other words, the seismic stream determines those elastic deformations and
stresses which are released in the earthquake sources during the earthquake.

It has been discovered that the earth's crust in the Gazli region is subject to
varied deformations—from the dominant tension (Tuzkoi bend, Lake Karakyr-2)
to a common slip (the greater part of the area, including the Gazli structure), and
the dominant compression (parts of the Bukhara-Ghissar fault) (Fig. 4).

The region consists of the blocks deformed in different directions. The directions
of predominant deformation are oriented across the existing faults. In the sublongitudinal direction the concealed Andakul fault is traced, defined by geophysical data
(TAL-VIRSKI *et al.*, 1984). Within the Gazli structure are distinguished volumes
with different directions of the predominant deformation: in the western part—lat-

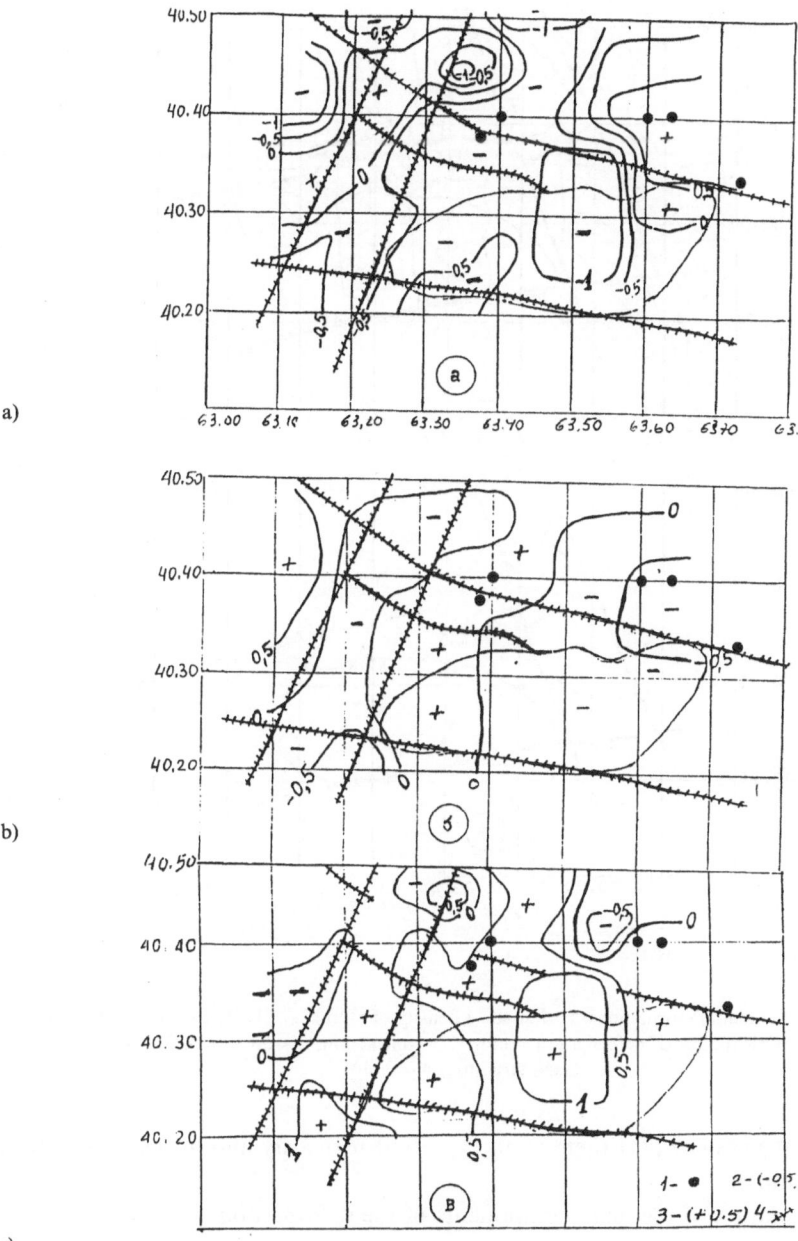

a)

b)

c)

Figure 4

Map of deformation velocity tensor components (unit 10^{-8} year^{-1}). Directions: a—latitudinal V_{xx}, b—longitudinal V_{yy}, c—vertical V_{zz}; 1—sources of the major shocks and aftershocks of the 1976 and 1984 earthquakes $6.3 < M < 7.3$; deformations: 2—(-0.5), deformations of contraction; 3—($+0.5$) deformation of dilatation, 4—faults.

a)

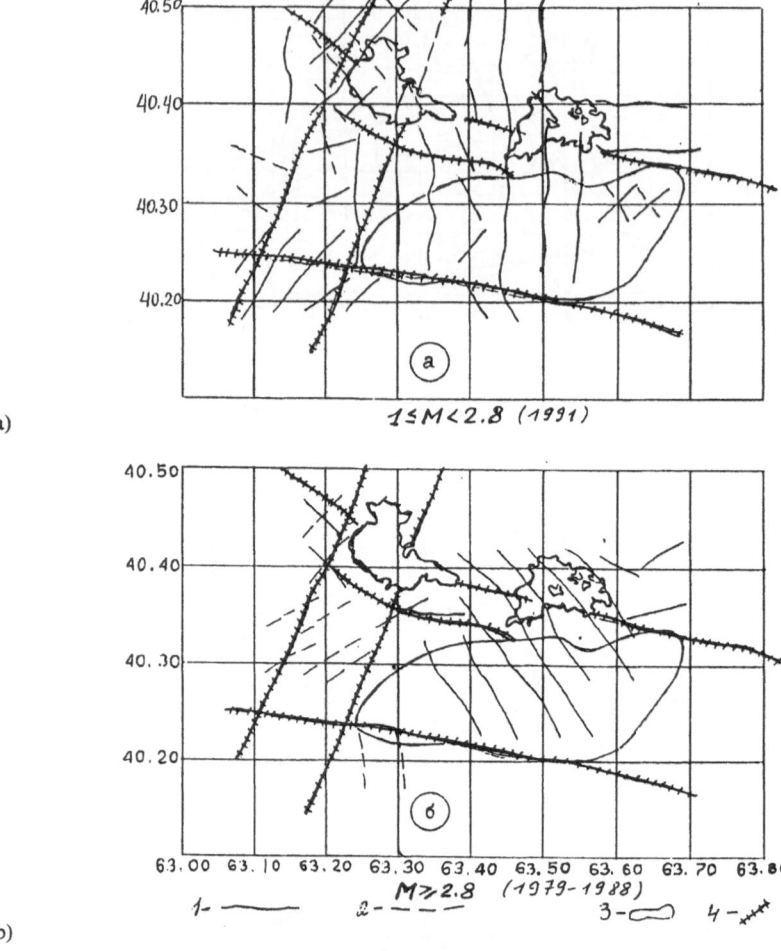

b)

Figure 5

Trajectories of the main deformations. Period and intensity of events. a—1991, 1 < M < 2.8; b—1979–1988, M > 2.8. Orientation of deformation vector: 1—contraction, 2—dilatation, 3—contours of the Gazli structure, 4—faults.

itudinal, in the eastern part there is no expressed regularity and in the central part it is longitudinal.

The calculation of the main components of the deformation velocity V_{ii} showed that in the latitudinal direction V_{xx} the earth's crust in the area is subject to the deformation of compression with the average velocity rate of $0.5 \cdot 10^{-8}$ per year (Fig. 5). The maximum value of compression deformation was observed within the central part of the Gazli structure $V_{xx} = 1.17 \cdot 10^{-8}$ per year and in the northwestern part of the Bukhara-Ghissar fault $V_{xx} = 1.0 \cdot 10^{-8}$ per year. The deformation of

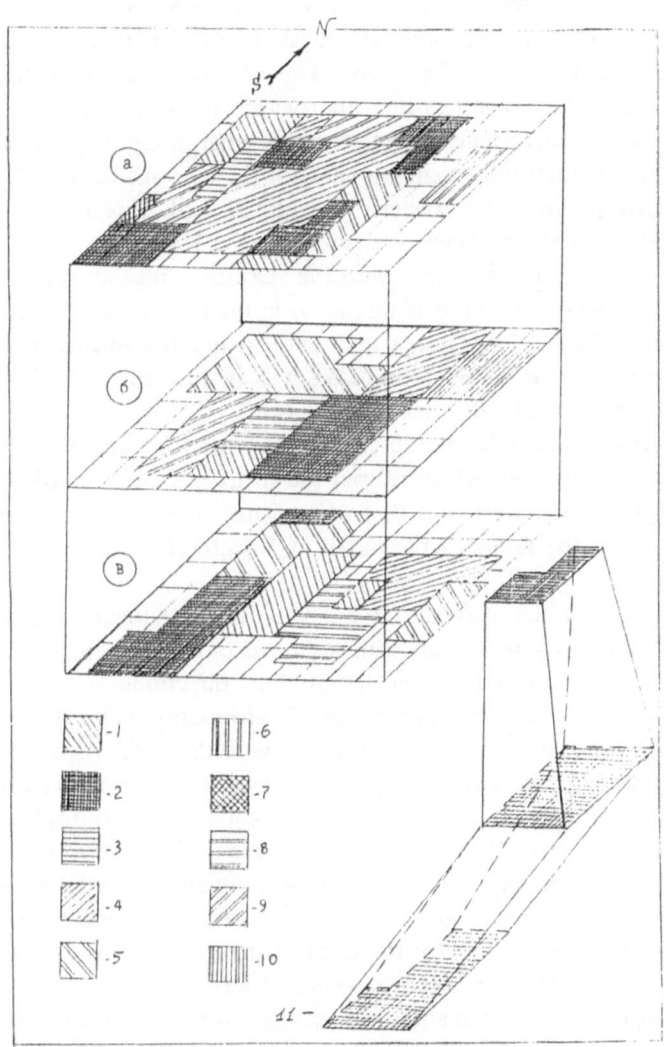

Figure 6

Types of deformations V_{zz} at separate depth intervals: a—1-4 km, b—4-7 km, c—7-10 km. 1—near-horizontal contraction northeast orientation with near-vertical dilatation; 2—the same, but the axis of near-horizonotal contraction of northwestern orientation; 3—near-horizontal dilatation of northeastern orientation with near-vertical contraction; 4—dislocation at sublongitudinal orientation of compression; the same with northwestern orientation of compression; 7—near-horizontal dilatation of northwestern orientation with contraction at the angle (30-60°) to the vertical: 8—the same, but with sublongitudinal orientation; 9, 10—dilatation and contraction at the angle (30-60°) to the vertical; 11—subvertical channel with increased seismicity and reduced viscosity in the interval 1-10 km.

tension was observed within the boundaries of Lake Karakyr-2, the eastern part of the Bukhara-Ghissar fault (within the focal zone of 04. 1976 earthquake) and the western board of the Tuzkoi depression. The velocity of deformation along the deep Bukhara-Ghissar fault was less than that within the Gazli structure.

The velocity of deformation of the earth's crust in the longitudinal direction is, on the whole, smaller: the maximum value of the tensions velocity $V_{yy} = 0.82 \cdot 10^{-8}$ per year corresponds to the western part of the Tuzkoi depression; the minimum value corresponds to the Gazli structure $V_{yy} = 0.3 \cdot 10^{-8}$ per year. The Bukhara-Ghissar fault is not homogeneous with the velocities and the types of deformation: in the longitudinal direction its main part is in the compression regime (the average velocity rate $0.15 \cdot 10^{-8}$ per year) and only within the focal zones of 05. 1976, $M = 7.0$ and 1984, $M = 7.3$ earthquakes it is in the tension regime (the average velocity rate $V_{yy} = 0.21 \cdot 10^{-8}$ per year).

The correlation of deformation velocities V_{xx} and V_{yy} can possibly be explained by the fact that the faults which restrict the structures under consideration are of a smaller depth in the west and east ($H = 5$ km) — Tuzkoi, Poperechny Karakyr, Rometan — than the faults in the north and south ($H = 15$ km), Gazli, Prodolny Karakyr.

The sources of the 1976 and 1984 $M > 7.0$ earthquakes and their strongest aftershocks $5.3 < M < 6$ coincide with the zones of diverse deformations (dilatation and contraction) in latitudinal and longitudinal directions.

The value of the vertical component of deformation velocity tensor testifies to the fact that the major part of the territory is in the uplift regime. The maximum velocity value of uplift $1.15 \cdot 10^{-8}$ per year is observed in the central part of the Gazli and the Utchkyr structures. This value corresponds to displacement velocity $V = 0.41$ mm per year. The subsidence zones are in the western part of the Tuzkoi depression, Lake Karakyr-2, and the eastern extremity of the Bukhara-Ghissar fault (focal zone of April 1976, $M = 7.2$, earthquake). The block of the earth's crust corresponding to the position of the Gazli structure, in terms of SDP, is expressed most brightly, and is most homogeneous and large.

The transition to the industrial regional gas storage regime led to the change of orientation in the directions of the main deformations and a more differentiated pattern of deformation of the earth's crust of the region.

The calculation of trajectories of the main deformations, those of dilatation and contraction for three depth intervals, 1–4, 4–7, 7–10 km, showed that the block of the earth's crust in the depth interval 1–4 km is differentiated the most (Fig. 6). The dimensions of the block's increase with the depth and the earth's crust becomes more homogeneous with the character of deformations. In the central part of the Gazli structure within the 1–10 km depth interval, a block, homogeneous by the character of deformations, is distinguished. This block coincides with the region of increased seismicity (PLOTNIKOVA *et al.*, 1994). This indicates the existence of a deep channel with low viscosity of the rock, by which migration of fluids may be taking place. This kind of gas accumulation mechanism, controlled by the subvertical zones of high cleft formation of basement rocks, is a geodynamic sign of the structures productivity.

The comparison of the vertical velocities of displacements due to seismic flow with recent vertical tectonic movements of the earth's crust (the levelling of 1st class) has revealed their direct relation and a high percentage (4%) of the seismic flow contribution to the tectonic one (in the Caucasus, 1%; Eastern Uzbekistan, 2%; Tadjikistan, 10%).

Discussion

The results obtained, testify that the active seismic processes in the Gazli region are connected not so much with the residual stress release in the focal zone of the earthquakes 05. 1976 and 1984, $M > 7.0$, but rather with the influence of the gas reservoir exploitation regime on the rocks with different rheologic properties.

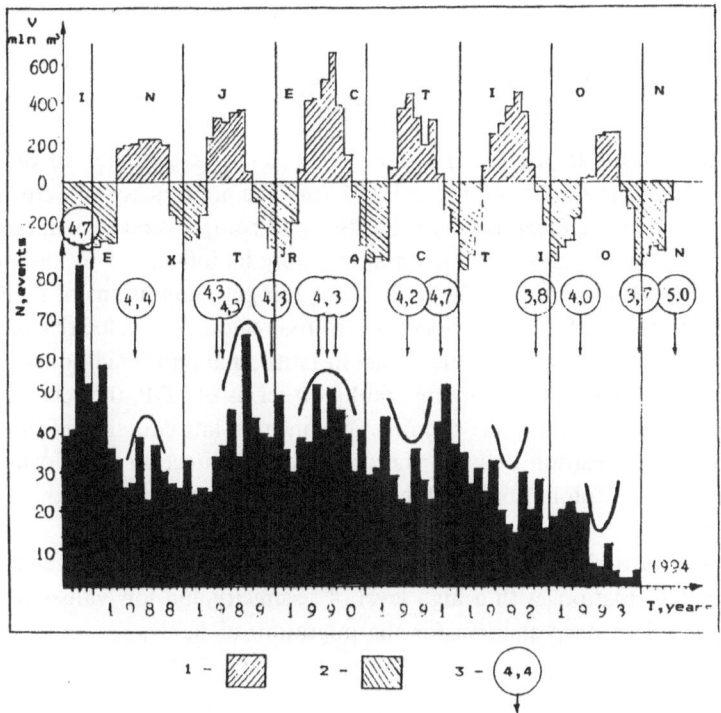

Figure 7

Seismic process development in Gazli region during the period of exploitation of the gas field as an industrial gas storage. Exploitation regime: 1—gas injection, 2—gas extraction, 3—appearance time of earthquakes $M > 4.0$.

The above conclusion is substantiated by the results of analyses of the course of the seismic process during the time period from 1988 (the beginning of the regional gas storage regime) to 1993, which included six cycles of gas injection-gas extraction (Fig. 7).

During the period of the first three cycles (1988–1990) gas extraction (reduction of the load on the earth's crust and seam pressure) was followed by a certain decrease in the earthquakes frequency of recurrence. Gas injection (increase of the load by $20 \cdot 10^3$ t and seam pressure from 7.2 to 11.7 atm.) is connected with the increase of seismic event numbers by 40–60%. The most severe $4.3 < M < 4.7$ earthquakes occurred after the change of the reservoir exploitation regime (the effect of pulsation stress) with the time delay 2.5–3.0 months (Fig. 7).

From 1991 to 1993 (4–6 cycles) the development of the seismic process in the region and the character of its dependence on the exploitation regime changes; the tendency of reduction in the number of shocks $M > 2.5$ against the background of the local minima is revealed. This change is likely to be due to the preparation of a strong earthquake $M = 5.0$ in the region in May 1994, which we predicted in 1990 on the basis of a statistical model of induced seismicity in the region (PLOTNIKOVA *et al.*, 1990).

Summary

The earth's crust in the Gazli region is subject to a complicated deformation process, especially in the depth interval 1–4 km. The main factor of deformation of the earth's crust in the Gazli region is background compression. A complex pattern of deformation was observed, which may possibly be influenced by the differences of the mechanical properties of the rocks, the geological conditions of their bedding and the orientation of faulting dislocations of the region. The velocities of deformations of block volumes of the earth's crust in latitudinal and longitudinal directions are controlled by the depth of tectonic faults. In terms of SDP, the relation between the velocities of tension and compression in the latitudinal and longitudinal directions and the relation between the predominant direction of deformation and the gas reservoir exploitation regime is most clearly manifest in the Gazli and Utchkir structures.

The subvertical block of the earth's crust is distinguished: homogeneous by the kind of the deformation, with a high level of seismicity and low values of viscosity, which is possibly a deep channel for the migration of fluids.

Acknowledgements

The authors acknowledge the support during the period of field works of the Université Joseph Fourier, Grenoble, and the staff of Gazlitransdobycha, Uzbekistan.

The research described in this publication was made possible in part by Grant MZEOOO from the International Science Foundation.

REFERENCES

GRASSO, J. R. (1990), *Hydrocarbon Extraction and Seismic Hazard Assessment*, EOS Trans. Am. Geophys. Union *71*, 1454.

NIKITIN, L.M., and YUNGA, S. L. (1977), *Methods of Theoretic Estimation of Tectonic Deformations and Stresses in Seismoactive Regions*, Izvestia AN SSSR, Fizika Zemli *11*, 54–67 (in Russian).

PLOTNIKOVA, L.M., FLYONOVA, M. G., and MAKHMUDOVA, V. I. (1990), *Induced Seismicity in the Gazli Gas Field Region*, Gerlands Beiträge zuy Geophysik, Leipzig *99*, 389–399.

PLOTNIKOVA, L.M., FLYONOVA, M. G., and IVANOVA, E. G. (1991), *Method for Determining the Depth of Faults in the Region of Source of Large Earthquake*, Certificate of authorship for the invention, N1684765, Moskow.

PLOTINKOVA, L. M. GRASSO, J. R., NURTAEV, B. S., and FRÉCHET, J. (1994), *Peculiarities of Seismic Process Development in the Gazli Region after 1976 and 1984 Earthquakes, M > 7.0*, Uzb. Geol. J. *2*, 11–15 (in Russian).

RIZNICHENKO, Yu. V. (1985), *Problems of Seismology*, Selected Works, M: Nauka 405 pp. (in Russian).

TAL-VIRSKY, B. B., KHUDAIBERGANOV, I. A., and PIVOVAROV, B. I. (1984), *The newest movements, geophysical fields and seismic process in the focal area*. In *Gazli Earthquakes of 1976*, M. Nauka, pp. 148–155 (in Russian).

YERKES, R. F., and CASTLE, R. O. (1976), *Seismicity and Faulting Attributable to Fluid Extraction*, Engin. Geol. *10*, 151–167.

(Received February 13, 1995, revised November 11, 1995, accepted December 11, 1995)

PAGEOPH, Vol. 147, No. 2 (1996)

0033–4553/96/020389–19$1.50 + 0.20/0

Analysis of Microtremors within the Provadia Region near a Salt Leaching Mine

P. KNOLL,[1] G. KOWALLE,[1] K. ROTHER,[1] B. SCHREIBER,[1] and I. PASKALEVA[2]

Abstract—By analysis of microtremors recorded with digital seismological monitoring equipment near the Provadia salt diapir (Bulgaria), two groups of events showing different characteristics have been detected in the vicinity of the salt production area. The first group of events has low magnitudes and is located at a distance of about 1 km from the top of the salt diapir. These events show low stress drops. The second group of tremors is located outside the salt diapir. The corresponding magnitudes and stress drops are larger. The first class of events seems to be related to processes at the contour of the salt leaching caverns, whereas the origin of the second group seems to be connected with stress redistribution processes around the salt body. Based on this analysis, the tectonic model of the Provadia salt diapir has been modified.

Key words: Monitoring of microearthquakes, salt diapir, seismotectonics of Provadia (Bulgaria).

1. Introduction

The study of induced seismicity includes an intense investigation concerning historical and recent natural seismicity of the area and of the adjoining regions. This prerequisite allows, after a sophisticated analysis, a division of the observed seismic events into natural and induced ones. Therefore, the creation of the corresponding data base must include two parts: the historical data file describing the seismicity of the region without anthropogenic influence, and the observational data dealing with the seismic regime of the studied area under conditions changing by human activity. Since the entire energetic spectrum of seismic events must be taken into consideration as well as the development of seismicity in time, the second task can be solved only by the operation of seismic monitoring stations or a corresponding network.

The salt mine under investigation is located in the Provadia region (43.06 N, 27.45 E) in the only known salt diapir of Bulgaria. Exploitation was started some

[1] GTU Ingenieurbüro Knoll, Potsdamer Str. 18A, 14513 Teltow, Germany.
[2] Bulgarian Academy of Sciences, Central Laboratory for Seismic Mechanics and Earthquake Engineering, Acad. Boncev str., block 3, 1113 Sofia, Bulgaria.

30 years ago by leaching caverns into the salt body. The Provadia region is characterized by low or moderate seismic activity with indications that seismicity has increased during recent decades. An intense discussion has taken place concerning the nature of seismicity in that area. A local net of accelerometers was installed some 15 years ago to study seismicity and ground shaking, and a set of records has been collected. However, the question concerning the natural or induced origin of the seismic activity of that region has not yet been solved.

In 1993 a German-Bulgarian collaboration between the Central Laboratory for Seismic Mechanics and Earthquake Engineering, Sofia and GTU, Teltow to study the seismicity in the Provadia region began. This project was supported by the Ministry of Science, Research and Culture of the State of Brandenburg, Germany.

2. The Seismotectonic State of the Provadia Region

The data base describing the seismotectonic state of the Provadia region is rather weak. Therefore only seismicity should be investigated within the framework of this report.

The greatest seismic hazard exists along the western and southern margins of the Balkan plate: the so-called Split seismic zone. The number of seismoactive faults was averaged in an area of 625 km² in this region. The highest values for Bulgaria are about $20 \, \text{km}^{-2}$ (Fig. 1). Neither the most seismic active fault zones of the Balkan plate nor the most seismic active intersections of fault zones are located in Bulgaria. Within the Provadia region the density of seismogenic faults is rather small $(0-2 \, \text{km}^{-2})$.

Based on the analysis of space images of the Bulgarian territory, GOCEV and MATOVA (1977) have established a link between photolineations, known tectonic structures and seismicity. They found that the Provadia region is in a zone with relatively high seismicity between 1900 and 1970, determined mainly by the Devnya fault in northern Provadia. BONCEV (1982) as well as VAPTSAROV and MISHEV (1982) analyzed the links between tectonics and morphology of Bulgaria with seismicity. The area around Provadia was found to be of moderate seismicity which originated in the fault structure of the "Diagonal Swell" (BONCEV, 1982).

A comprehensive seismic zoning map of Bulgaria was compiled by BONCEV *et al.* (1982), based on geological and geophysical data. The Provadia region belongs to the area of moderate seismicity of Bulgaria. The relative seismic danger of Provadia is about half of that of the Struma valley and the expected maximum magnitude reaches 4–5. Within a period of 1,000 years a seismic intensity of VII (MSK) is expected, and for 10,000 years the intensity of VIII (MSK) should be taken into consideration.

DOTZEV and YUNGA (1989) determined compressional stresses in a northeast-southwest direction for the northern part of Bulgaria by means of fault plane

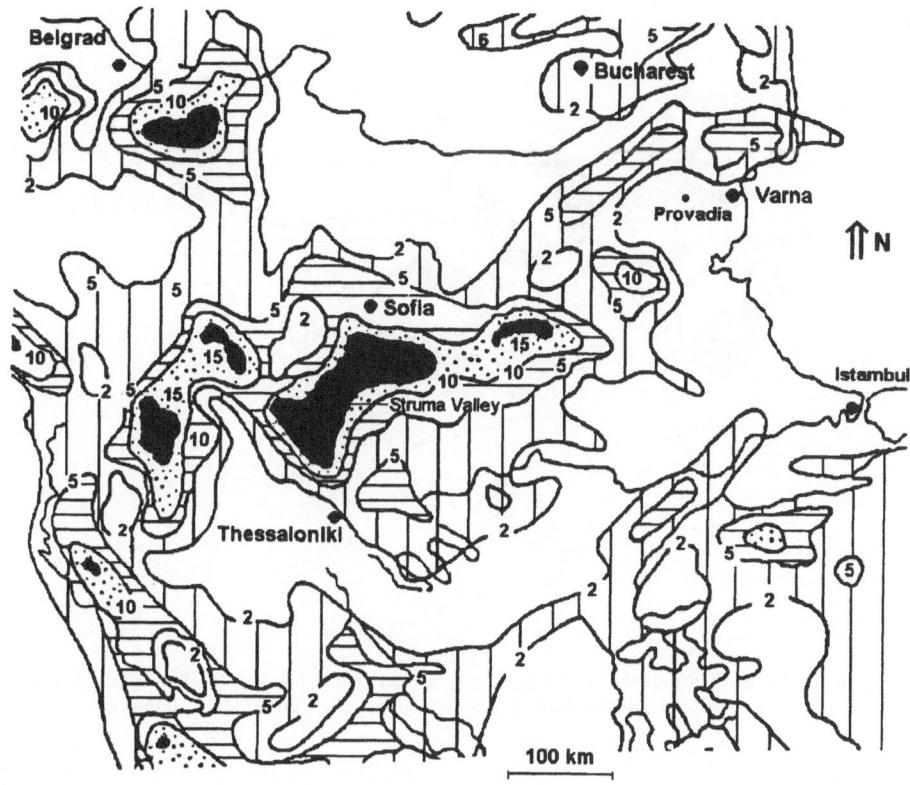

Figure 1
Density of seismogenic faults of the Balkan region (after DOBREV and SCHUKIN, 1984). It is clearly
visible that the Provadia region is characterized by a rather low fault density (0–2 km²).

solution studies, as well as SHANOV *et al.* (1988) who detected this direction
seismologically and by means of the dispersion of shear joints in Upper Pliocene
and Lower Cretaceous limestones of Northern Bulgaria.

SHANOV (1990) confirmed this direction by reconstructing the tectonic stress
field in northeastern Bulgaria, based on the tectonic fracture pattern and physical
anisotropy of rocks from the Lower Cretaceous period up to the Upper Pliocene.
A clockwise rotation of the maximum and minimum compressional axes was found.
The maximum compression after the Lower Cretaceous period trends NE-SW,
whereas after the Pliocene and compression is directed NW-SE.

The direction of compressional stresses within the Provadia region was analyzed
by GEORGIEV (1990) by means of focal mechanisms of five earthquakes which
occurred in this region. The focal coordinates determined by the NEIS (US
Geological Service) were used in this study. It can be stated that the general stress
pattern of the Provadia region agrees with that of the northern part of Bulgaria.

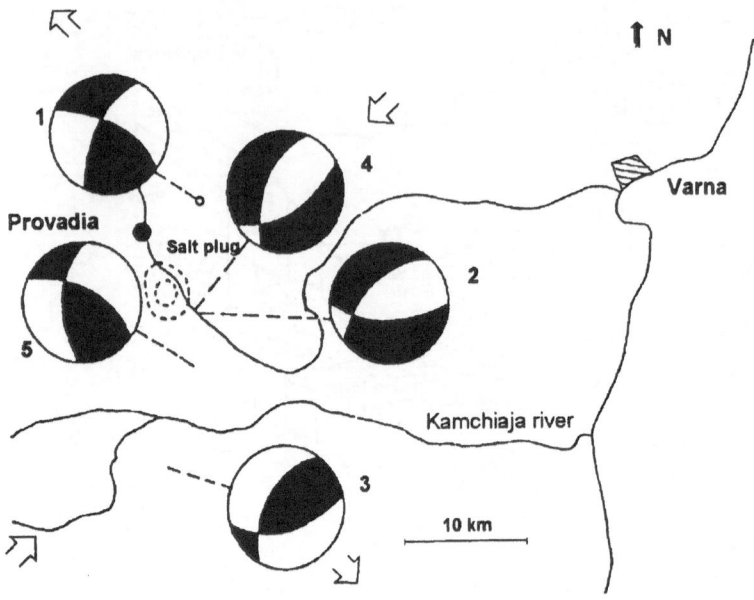

Figure 2
Source mechanisms of earthquakes with magnitudes greater than 3.6 within the Provadia region (after
GEORGIEV, 1990). Clearly the Provadia salt diapir normal faulting is the dominant one.

The compressional stresses are directed northeast-southwest. It was found that two
earthquakes near the salt body (below it) show normal faulting (Fig. 2). However,
it should be mentioned that some discrepancies appeared when comparing the fault
plane solutions of GEORGIEV (1990) with those of SOLAKOV and SIMEONOVA
(1993). Most of the focal mechanisms determined by the Bulgarian Seismological
Network (BSN) have been characterized as fairly poor (SOLAKOV and SIMEONOVA,
1993).

A general scheme of the tectonic state of the Provadia region has been presented
by PASKALEVA *et al.* (1994). Within this scheme the seismicity shows a distinct
pattern which is clearly related to regional tectonics and mining activity. Three
different zones of seismic activity around the salt diapir were postulated. Zone I is
without seismic activity. It contains an area of about 4–5 km in diameter. Zone II
is situated outside zone I with a radius of about 10 km. Within this zone earthquakes with a magnitude up to 4 have been observed. Zone III contains the zone
of background activity outside Zone II. It represents the regional tectonic condition. This model is a rather qualitative one since the location capability and
accuracy are fairly poor for this region.

Although the general features of the stress pattern in northeastern Bulgaria are
known, *in situ* determinations of stresses within the Provadia region and around the

Table 1

Microseismic focal parameters for earthquakes of the time period 1981–1990 felt at Provadia

Date	Coordinates	Depth (km)	D (km)	M	I_0 (MSK)	I_{Prov} (MSK)
1981/07/23	43.15 N, 27.45 E	4	10	3.9	VI–VII	IV
1981/09/04	43.15 N, 27.5 E	8	7	3.5	V–VI	IV–V
1983/11/10	43.2 N, 27.5 E	6.5	45	4.5	VI–VII	III
1885/06/12	43.2 N, 27.45 E	7	15	4	VI	IV
1986/06/12	43.2 N, 26.0 E	13	90	5.7	VIII	IV

salt body would improve significantly knowledge regarding the tectonic regime of the studied geological structure and could be very helpful in estimating the anthropogenic factor due to leaching.

The main problem in analyzing the seismicity of Bulgaria is the lack of data for nearly 5 centuries. Consequently, there is no possibility of performing a long-term analysis of the seismicity of the region, and neither the *b* value of the Gutenberg-Richter-relation nor the maximum possible magnitude for earthquakes in Bulgaria could be determined with a sufficient degree of accuracy, especially for the Provadia

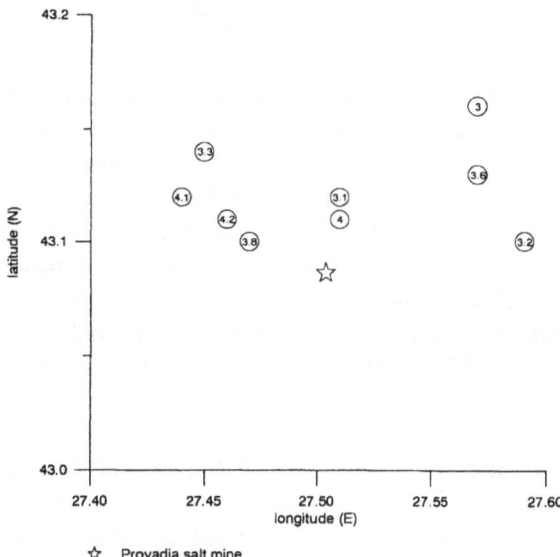

Figure 3
Distribution of earthquake foci for the time period 1981–1990 within the Provadia region. The numbers inside the circles represent the magnitude. Data are taken from the Bulgarian bulletin (SOLAKOV and SIMEONOVA, 1993).

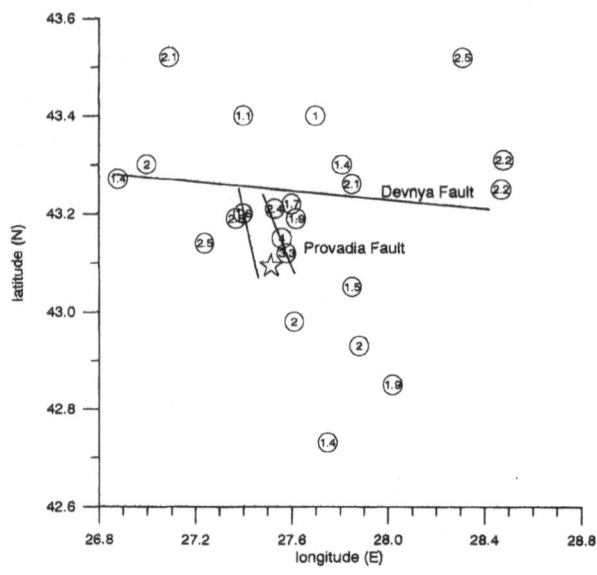

Figure 4
Distribution of earthquake foci for the time period January 1993–April 1994 within the Provadia region.
The numbers inside the circles represent the magnitude. Data are taken from the Preliminary Bulgarian
bulletin (RANGELOV, 1994).

region. From 1981 GLAVCHEVA (1993) compiled an atlas of isoseismal maps of
Bulgaria. It contains four isoseismal maps of earthquakes near Provadia. The
parameters of these events are given in Table 1, together with those of a more
distant earthquake which was strongly felt in Provadia, too.

M denotes the magnitude of the event, D the epicentral distance to the Provadia
salt diapir, I_0 the seismic intensity within the epicentral region, and I_{Prov} the
intensity at the Provadia salt mine site. It follows from the above Table that within
the Provadia region the maximum observed intensity is about IV–V degrees MSK
for this ten-year period. The epicentral distance of about 10 km and more makes it
very unlikely that these events were influenced by the mining activity at the
Provadia salt body.

By means of the Bulgarian Seismological Network (BSN) several events have
been located within the Provadia region (SOLAKOV and SIMEONOVA, 1993;
RANGELOV, 1994). Figures 3 and 4 present the distribution of earthquake foci
within the Provadia regions for the time periods 1981–1990 and January 1993–
April 1994, respectively.

The distribution of earthquakes with magnitudes greater than 3.0 for 1981–1990
shows linear trends, indicating a close correlation to the course of the Devnya fault

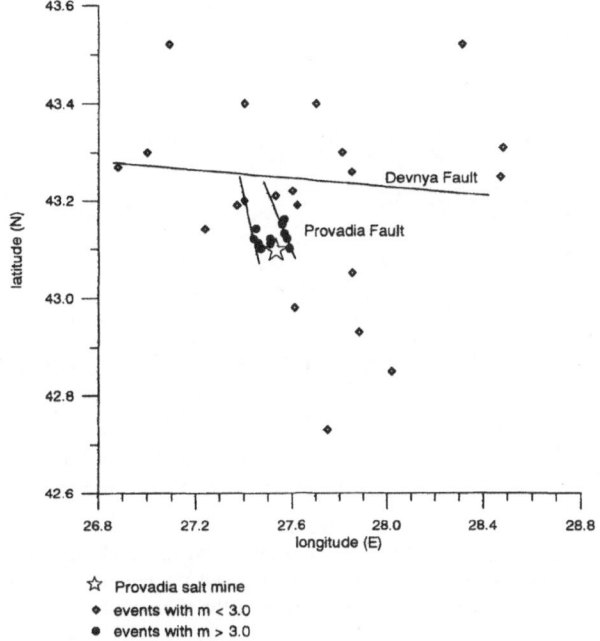

Figure 5
Distribution of earthquake foci for the time periods 1981–1990 and January 1993–April 1994 within the
Provadia region. Dots represent events with $m < 3.0$ and diamonds $m > 3.0$.

striking north of the salt body as well as to the submeridional directed Provadia
fault (GOCEV and MATOVA, 1977; GEORGIEV, 1990; VELICHKOVA and SOKEROVA,
1980). The events of the second period are distributed in a larger area, however the
strongest ones (magnitudes 3.3 and 4.0) are located very close to those of the
previous time period.

When merging the two data sets several linear structures could be identified
which are parallel to the directions of the main fault systems in the northeastern
part of Bulgaria. Clearly visible are the subequatorial striking and the perpendicular
one (Fig. 5). It should be mentioned additionally that the stronger events are
located in the vicinity of the salt body, where a crossing of the subequatorial and
submeridional fault systems is assumed.

3. Seismic Monitoring of the Provadia Salt Mining District

Until 1994 the only instrument which existed for monitoring seismic activity
within the Provadia region was the National Bulgarian Seismological Network

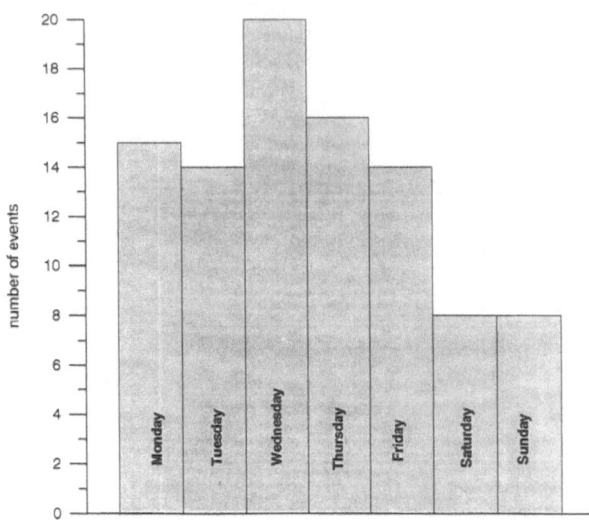

Figure 6
Distribution of triggered events by the REFTEK equipment at Provadia. The number of triggered events
is not equal for all the days of the week. During the weekend the number is smaller, indicating that
certain events must be caused by human activity in the vicinity of the recording point.

(BSN) with the nearest station Provadia which is operating experimentally to date, and a network of accelerographs with a rather high trigger level. By these means no possibility had existed for monitoring very weak events near the salt diapir.

To improve the experimental base for observing microseismic activity in this region, seismological equipment containing a REFTEK-data acquisition system 72A–07 (DAS), a GPS timing system 111A, and a three-component accelerometer SSA 320 has been provided by GTU, Teltow. The equipment was installed in an administrative building of the mine GEOSOL at the top of the salt diapir. The continuous data recording started on June 1, 1994. The data have been periodically read out and provided for further analysis.

Ninety-five events were recorded during the first three months. The maximum number of events per day was eight. Figure 6 shows the daily distribution of events throughout the week, and Figure 7 presents their distribution during the recording time (24 hours). From these figures it is clearly recognizable that most of the events occurred on normal working days (Monday through Friday) and during day time. Only a few events were recorded at night. This indicates that most of the recorded events were induced by human activities near the recording site. The same kind of distributions could be found for the next observation period.

The noise ratio for the Provadia site is about 12 dB, not taking into account noise bursts due to traffic or technological processes. For technologically induced

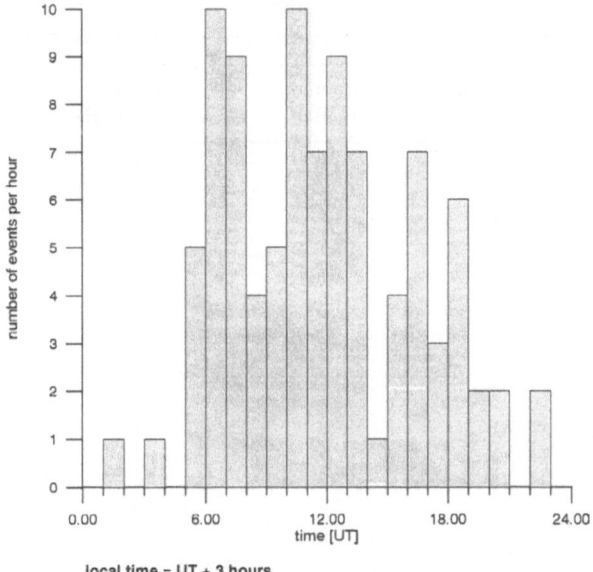

local time = UT + 3 hours

Figure 7
Time distribution of triggered events over 24 hours. The amount of triggered events at night is
significantly smaller than during daytime.

noise bursts an amplitude ratio of the noise disturbances to the undisturbed ground
noise of about 35 dB has been observed.

By comparing the records of the first three months with master events, 12 of
them were identified as containing possible microearthquakes, meaning a false
alarm ratio of 87%. This ratio seems to be high, nevertheless it has been decided not
to change the trigger value for providing the possibility of detecting very weak
events near the salt deposit.

Within the second three months period 26 events were recorded. These files
contain 11 records of near microearthquakes. It is obvious that the number of
possible microearthquakes recorded during both periods of time is similar. However
the number of triggered records of the second period is significantly smaller. If the
events identified as possible microearthquakes are indeed quakes this would imply
that the seismogenic regime has not changed. In this case the false alarm rate is
about 58%. The distributions of false triggerings in time indicate that beyond
September the traffic activities near the salt mine have decreased significantly.

The main tasks of seismic monitoring are the determination of hypocentral
coordinates and the energy of recorded events. Since no other high-sensitive station
exists near the salt diapir of Provadia, only a single station data is available for
analysis. Therefore the distance can only be determined by means of the onset time

Table 2

Characteristics of microearthquakes in the Provadia region recorded by the REFTEK monitoring system

Date	Epicentral distance (km)	Magnitude M_D(Bul)	Corner frequency f_0(Hz)	Remarks
06/20/94	4.5	3.1	5	
07/01/94	5.1	2.5	8	
07/06/94	3.7	3.0	6	
07/12/94	1.5	2.1	8	
07/21/94	1.0	1.7	8	*
07/25/94	0.9	1.6	10	*
07/25/94	1.1	1.7	9	*
07/26/94	1.0	1.6	7	*
07/27/94	0.6	1.4	7	*
08/11/94	5.5	2.9	7	
08/18/94	7.1	2.8	8	
09/12/94	5.5	2.6	5	
09/25/94	3.5	2.8	10	
09/27/94	4	2.8	5	**
09/27/94	4	2.5	7	
09/28/94	4	2.5	8	
10/06/94	3	3.1	3	
10/06/94	0.8	1.2	12	*
10/07/94	0.5	1.2	12(6)	* possibly 4 events
10/09/94	4	2.9	11	
10/09/94	4	2.5	7	
10/25/94	4	2.8	6	
11/18/94	4	2.5	7	

* possible swarm.
** There exists a second interpretation: The record contains two microearthquakes in a distance of $D \approx 1$ km with magnitudes of M_D(Bul) = 1.4.

difference dt of the record phases travelling with different velocities v_1 and v_2. The distance D (in km) between source and receiver can be determined by

$$D = dt(v_1 - v_2)/(v_1 v_2) = (t_2 - t_1)(v_1 - v_2)/(v_1 v_2) = (t_2 - t_1)/v^*$$

where t_1 and t_2 are the respective onset times of phases 1 and 2. To date there are no concrete travel time curves available for the Provadia region. Thus, the standard values of velocities of seismic waves within the earth's crust $v_1 = v_P = 6.0$ km sec^{-1} and $v_2 = v_S = 3.5$ km sec^{-1} have been applied. For further analysis the value $v^* = 8.1$ km sec^{-1} has been used. This value may cause some systematic errors of the distances determined. The results of the analysis are given in Table 2. The column M_D(Bul) gives the magnitude values computed by means of the Bulgarian formula (SOLAKOV and SIMEONOVA, 1993) used for compiling the Bulgarian seismological bulletin:

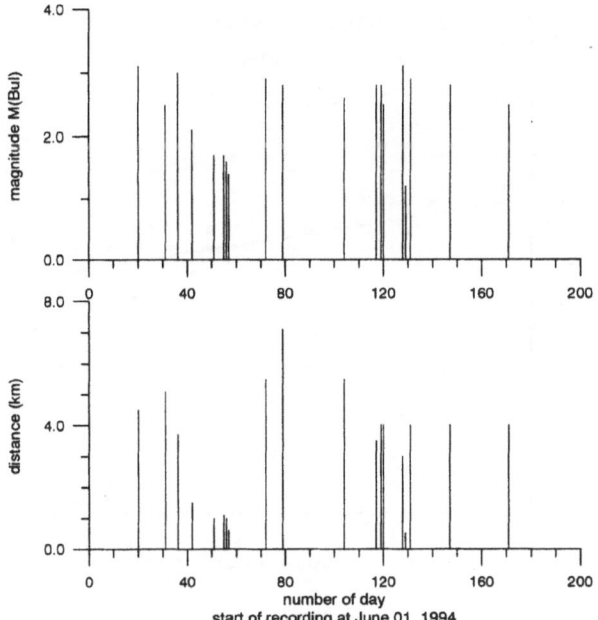

Figure 8
Magnitude-time and epicentral distance-time distributions of Provadia events recorded by the REFTEK
equipment.

$$M_D(\text{Bul}) = 1.98 + 2.72 \log \tau - 0.00023D \text{ (km)},$$

where τ is the duration of the seismic record.

To determine the magnitude of the events, the duration magnitude formula has been used. The coefficients of the formula are mean values changing from one tectonic unit to another. In a further step of investigation it would be very useful to establish a local magnitude scale for the Provadia region.

In Table 2 two groups of events having nearly the same epicentral distances of about 4 km and of about 1 km can be observed. The nearer events are weaker and cluster in time. Due to very narrow clustering in space (distance) and time, these events spanning the time periods July 21–July 27 and October 06–October 07 have been characterized as possible events of a small "swarm" in the very vicinity of the recording location. The magnitudes of these events are also within the range $M_D(\text{Bul}) < 1.7$. The second group of events is located in a distance range of about 4 km. The magnitudes of this group are generally about one magnitude unit larger. Figure 8 displays the distribution of microearthquakes for the entire observation period (June 01–November 26, 1994).

Since the events are very weak and only one sensitive station collects data, there is no possibility of determining the epicentral coordinates. Therefore it is unknown

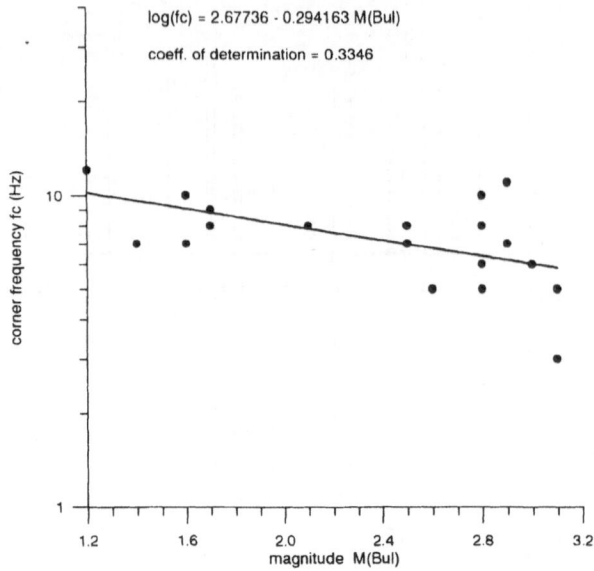

Figure 9
Dependence of the corner frequency on the magnitude. Data recorded by the REFTEK equipment
within the Provadia region have been used.

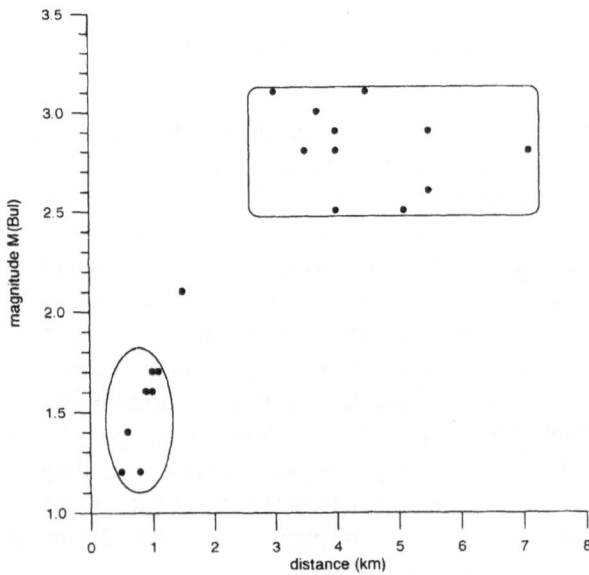

Figure 10
Distribution of events in the magnitude-distance space.

where the foci have been located, nonetheless it should be stated that the deter-mined epicentral distance is in the range of the depth of the production caverns. Furthermore, there is no knowledge of possible special technological activities within the time period of the "swarm".

The installation of further highly sensitive seismic sensors in the vicinity of the salt body would be an important step towards increasing the knowledge of the tectonic seismogenic behavior of the Mirovo salt body and its interaction with the technological activity while leaching.

The corner frequencies are generally in the expected range, but due to the weakness of the events a rather large scattering is observed. Figure 9 illustrates the dependence of the corner frequency f_0 on the magnitude M_D(Bul). A regression curve is additionally shown. The coefficient of determination is rather small, indicating a large scatter of the data. To compare this dependence with the above given formula, an exponential curve fit was performed, leading to the following functional dependence:

$$\log f_0 = 2.67736 - 0.2941 M_D\,(\mathrm{Bul}).$$

The coefficient of determination is equal to 0.3346. Within the investigated magni-tude range the exponential fit differs slightly from the linear one. The above dependence of the corner frequency f_0 on the magnitude M_D(Bul) for events in the Provadia region is quite similar to that based on California data.

Figure 10 depicts the clustering of recorded events in the magnitude-distance space. Two separate groups of events are clearly recognizable. One group is characterized by small magnitudes and small epicentral distances whereas the second group has larger magnitudes and epicentral distances. Additionally, the first group exhibits a strong clustering in time (please, compare with Fig. 8).

From the above figures it is obvious that weak events tend to have small epicentral distance and *vice versa* and to cluster in time. Comparing the events recorded by the REFTEK-equipment with those recorded by the Bulgarian Seismo-

Table 3

Events reported by the Bulgarian Seismological Network (BSN)

Date	Source	Depth (km)	M_D(Bul)
02/06/94	43.35 N; 27.64 E	0	1.0
04/15/94	42.73 N; 27.75 E	10	1.4
05/27/94	43.08 N; 27.75 E	12	2.2
08/07/94	42.47 N; 26.32 E	9	2.4
08/18/94	43.04 N; 27.30 E	0	2.5
08/23/94	43.35 N; 27.56 E	1	2.5
08/30/94	43.32 N; 28.22 E	6	2.3
11/04/94	43.45 N; 27.76 E	0	2.2
11/10/94	42.94 N; 26.99 E	20	2.0

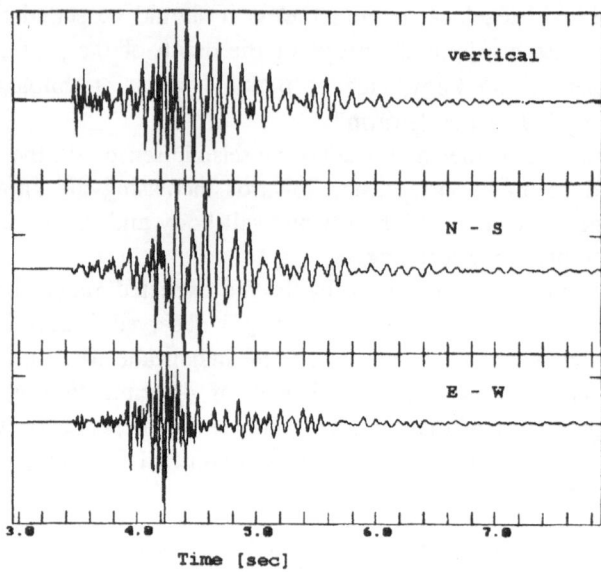

Figure 11
Record of the microearthquake of August 18, 1994 within the Provadia region.

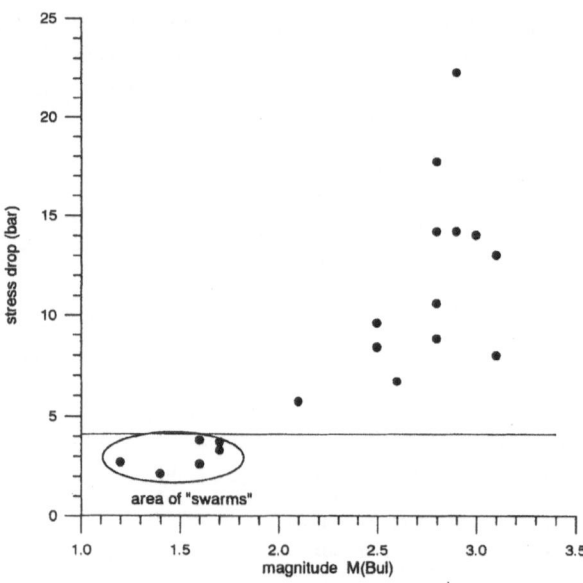

Figure 12
Dependence of the stress drop $\Delta\sigma$ on the magnitude for microearthquakes at Provadia. Clearly recognizable are two areas of clustering. Smaller events ($M(\text{Bul}) < 2.0$) tend to cluster into the "area of swarms".

logical Network (BSN) only one event recorded by both means (08/18/94) could be identified. For 1994 the BSN has determined the following events within the Provadia region (Table 3).

The event of August 18, 1994 (Fig. 11) was felt in Varna, but due to its small magnitude it was not reported by the PED (Preliminary Epicenter Determination of the NEIS/USGS). Until now other agencies have not reported this event. A more detailed analysis of this event reveals a corner frequency of the P wave of about 8 Hz.

Using the collected data of Provadia events and applying the well-known seismological scaling laws for magnitude (seismic moment) and corner frequency, it will be possible to determine some seismotectonic parameters of the source (EVERN-DEN, 1975; GROSSER and HURTIG, 1980; HURTIG and STILLER, 1984). Applying the circular source model of Madariaga, the source parameters source radius r in km, stress drop $\Delta\sigma$ in bar, and dislocation δ at the fault plane in cm were estimated.

Table 4 shows that the dislocations are smaller than 1 cm. The sources have dimensions of about 100 m, determined by means of the corner frequency data and the circular source model. Stress drop values vary between 2 and 23 bars. Those stress drops normally prove shallow seismic events.

Table 4

Source parameters of events in Provadia

Date	r (km)	$\Delta\sigma$ (bar)	δ (cm)
06/20/94	0.185	13.1	0.70
07/01/94	0.115	9.6	0.32
07/06/94	0.154	14.0	0.62
07/12/94	0.115	5.7	0.19
07/21/94	0.115	3.3	0.11
07/25/94	0.092	3.8	0.10
07/25/94	0.103	3.7	0.11
07/26/94	0.132	2.6	0.10
07/27/94	0.132	2.1	0.08
08/11/94	0.132	14.2	0.54
08/18/94	0.115	14.2	0.47
09/12/94	0.185	6.7	0.36
09/25/94	0.092	17.7	0.47
09/27/94	0.185	8.8	0.47
09/27/94	0.132	8.4	0.32
09/28/94	0.115	9.6	0.32
10/06/94	0.308	8.0	0.70
10/06/94	0.077	2.7	0.06
10/07/94	0.077	2.7	0.06
10/09/94	0.084	22.3	0.54
10/09/94	0.132	8.4	0.32
10/25/94	0.154	10.6	0.47
11/18/94	0.132	8.4	0.32

For all events in the vicinity of the recording point the stress drop $\Delta\sigma$ is rather small ($\Delta\sigma < 4$ bar), as well as the dislocation. This may be an indication that these events are connected with processes around or within the salt body. The dependence of the stress drop $\Delta\sigma$ on the magnitude M_D(Bul) is shown in Figure 12. Clustering of the "swarm"-like events is clearly visible, indicating a possible different generation mechanism of these events. Unfortunately it has been impossible to construct fault plane solutions due to the lack of sensitive stations within the Provadia region.

The recording of about half a year in the vicinity of the salt body at Provadia has shown that there are possibly two types of microtremors clustering within distance of the salt diapir as well as in time. An initial group consists of weak events with magnitudes M_D(Bul) < 1.8 occurring in small "swarms" and which are possibly related to stress redistributions in the vicinity of the diapir. The stronger events occur at greater distance to the diapir and show a link to the tectonic fault system around the salt diapir.

4. Discussion

Referring to KUHNT et al. (1989) the stress drop is very sensitive to the type of event. Small stress drops indicate induced seismic events caused by destruction of pillar edges and parts of the contour of underground openings, whereas large stress drops are mostly connected with tectonic components and stress redistributions in the rock massif surrounding the mine. Applying these findings together with the results of the seismic data analysis to the situation at the Provadia salt mine, it can be stated that the two types of events represent two different generation mechanisms of seismic events.

The first mechanism consists of possible rockfalls within the caverns and generates small events with low stress drop. The stress drop of these events should be small due to the very limited rock volume involved in those processes. This mechanism has been confirmed by numerical modelling by use of three-dimensional finite-element models of the cavern system in the salt diapir of Provadia (KNOLL et al., 1995).

The second mechanism represents stress redistributions due to mining and fracturing in the rock massif near the salt diapir. The finite-element modelling allowed the conclusion that stress concentrations within the salt body dissipate very rapidly with distance. Therefore the gap of seismic foci around the production area to the outer border of the diapir could be found. Stress changes due to salt production and the tectonics of the Provadia region may cause small tremors around the diapir. Therefore, the tectonic model of the Provadia salt diapir region (PASKALEVA et al., 1994) should be modified as follows (Fig. 13):

● the proposed Zone I where no seismic events have occurred should be
  subdivided into the Zones 1a and 1b,

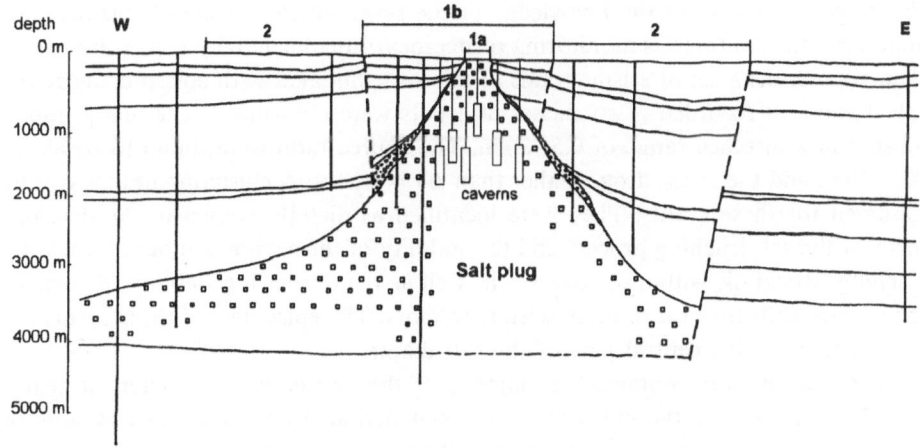

Figure 13
Seismotectonic model of the Provadia salt diapir region.

- Zone 1a contains the cavern field within the salt diapir,
- Zone 1b contains the outer part of the salt diapir,
- the former Zone II, which has been explained as the zone with small seismic events caused by mining activities, is transformed into Zone 2 and is located in the vicinity of the salt diapir.

The analysis demonstrates that weak seismic events have been generated within Zone 1a due to rockfalls or similar mechanisms. Within Zone 1b there are no seismic events due to the creeping behavior of the salt rock. Zone 2 is a highly tectonically dropped region near the diapir. Within this zone are located the Provadia fault as well as other local faults. The interaction of regional tectonic stresses and those mining triggered has changed the stress distribution within this zone such that small events may be generated here.

One of the most important problems concerning investigations of the seismicity in the Provadia region is the accurate determination of the focal coordinates. It can be solved only by installation of further highly sensitive seismic monitoring stations near the salt diapir. With such observations it will be possible to correlate the generation mechanism of seismic events with the local and regional tectonics as well as with technological activities within the mine.

5. Conclusions

The installation of a high resolution REFTEK seismological monitoring station in the lower floor of an administrative building near the top of the salt diapir of

Provadia has improved the knowledge of the seismotectonic state of this area. In June 1994 the continuous monitoring of microearthquake activity began. For more than six months a set of seismograms of microearthquakes with epicenters near the salt diapir was recorded. Two classes of events were recognized. The first group is located in a distance range of 0.5–2 km. The source radii were found to be about 80–120 m and the stress drop smaller than $\Delta\sigma = 4$ bars. A clustering in time can be observed for these events. They were identified as directly related to the development of the salt leaching process and the stability of the cavern contours formed by leaching. Rockfalls within the caverns as well as the decreasing stability of parts of the cavern walls may cause small seismic tremors. The epicenters of these events are concentrated in the central part of the salt diapir.

The second class contains the majority of the events and is distributed nearly equally in time over the entire observational period. The epicentral distances of these events are in the range of about 4–8 km. These events are characterized by magnitudes larger than 2, source radii larger than 100 m, and stress drops greater than $\Delta\sigma = 8$ bars. The interaction of mining induced stresses and the regional tectonic stresses are assumed to cause stress concentrations on existing faults and to generate small seismic events.

Spatially both groups of events are separated by a gap representing the outer part of the diapir. The gap is caused by the creeping behavior of the salt.

The study of recent seismicity, especially within the context of induced events, requires the installation of highly sensitive seismic equipment. Currently there exists only one digital station with high resolution power within the Provadia region. Therefore no fault plane solutions of weak events could be determined. The combination of the described approach of determining stress drops together with fault plane solutions would significantly improve the possibility of specifying the nature of seismic events in the vicinity of the Provadia salt diapir.

Acknowledgement

The authors wish to express thanks to the administration of the GEOSOL, Provadia, for its kind assistance, especially for allowing the installation of the REFTEK equipment in the administrative building of GEOSOL.

REFERENCES

BONCEV, E. (1982), *Seismotectonic Features of Bulgaria*, Geolog. Balcan. *12* (2), 71–98.
BONCEV, E., BUNE, V. I., CHRISTOSKOV, L., KARAGJULEVA, J., KOSTADINOV, V., REISNER, G. J., RISHIKOVA, S., SHEBALIN, N. V., SHOLPO, V. N., and SOKEROVA, A. (1982), *A Method for Compilation of Seismic Zoning Prognostic Maps for the Territory of Bulgaria*, Geol. Balcan. *12* (2), 3–48.

DOBREV, T. B., and SCHUKIN, J. K. (1984), *On the Problems of Structure, Geodynamics, and Seismicity of the Lithosphere in the Carpathian-Balkan Region*, Geotect., Tectonophys. and Geodyn. *17*, 35–59 (in Bulgarian).

DOTZEV, N., and YUNGA, S. (1989), *Fault-plane Solutions and Seismotectonic Deformations Study for the Central Balkan Region*, Proc. XXI Gen. Assem. ESC, Sofia, 271–277.

EVERNDEN, J. F. (1975), *Seismic Intensities of Earthquakes and Related Parameters*, Bull. Seismol. Soc. Am. *65*, 1287–1313.

GEORGIEV, Tz. (1990), *Contemporary Field of the Tectonic Stresses and Dislocation Process in the Region of the Mirovo Salt Body*, Bulg. Geophys. J. *XVI* (4), 63–70 (in Bulgarian).

GLAVCHEVA, R. (1993), *Atlas of Isoseismal Maps*, Geophys. Inst. BAS, Sofia, 67 pp.

GOCEV, P. M., and MATOVA, M. (1977), *The Present Fault Mosaic in Bulgaria and the Seismic Activity*, Geotect., Tectonophys. and Geodyn. *6*, 32–47 (in Bulgarian).

GROSSER, H., and HURTIG, E. (1980), *On the Analysis of the Fracture Process by Seismological Means*, Z. geol. Wiss. *8*, 303–314 (in German).

HURTIG, E., and STILLER, H., *Earthquakes and Earthquake Hazard* (Akademie-Verlag, Berlin 1984) 328 pp. (in German).

KNOLL, P., SCHREIBER, B., KOWALLE, G., ROTHER, K., PASKALEVA, I., and KOUTEVA, M. (1995), *Analysis of Dynamic Stability of a System of Caverns in the Salt Diapir of Provadia/Bulgaria*, Proc. 8th Int. Congr. Rock Mechanics, Tokyo, Sept. 25–29 (in press).

KUHNT, W., KNOLL, P., GROSSER, H., and BEHRENS, H.-J. (1989), *Seismological models for induced seismic events*. In *Seismicity in Mines* (Gibowicz, S. J. ed.) Pure and Appl. Geophys. *129* (3/4), 513–521.

PASKALEVA, I., MANEV, G., and KOUTEVA, M. (1994), *Induced Seismicity at Mirovo Salt Deposit, Bulgaria*, unpubl. manuscript.

RANGELOV, B. (1994), Personal communication.

SHANOV, S. (1990), *Tectonic Stress Field in Northwestern Bulgaria*, Geolog. Balcan. *20* (4), 37–47.

SHANOV, S., GEORGIEV, Tz., and DIMITROV, B. (1988), *Contemporary Tectonic Stress Field in Northwestern Bulgaria*, Rev. Bulg. Geol. Soc. *XLIX*, part 1, 39–46 (in Bulgarian).

SOLAKOV, D. E., and SIMEONOVA, S. D. (ed.) (1993), *Bulgaria — Catalogue of Earthquakes 1981–1990*. Geophys. Inst. BAS, Sofia, 39 pp.

VAPTSAROV, I., and MISHEV, K. (1982), *Main Features and Dynamics of the Morphostructures in Bulgaria: Significance for Seismic Zoning*, Geolog. Balcan. *12* (2), 99–116.

VELICHKOVA, S., and SOKEROVA, D. (1980), *Analysis of Seismic Events in Bulgaria*, Bulg. Geophys. J. *VI*, 58–72 (in Bulgarian).

(Received December 12, 1994, revised September 20, 1995, accepted September 25, 1995)

PAGEOPH, Vol. 147, No. 2 (1996)

0033-4553/96/020409-10$1.50 + 0.20/0

Induced Seismicity at Wujiangdu Reservoir, China: A Case Induced in the Karst Area

Hu Yuliang,[1] Liu Zuyuan,[1] Yang Qingyuan,[1] Chen Xiancheng,[1] Hu Ping,[1] Ma Wentao,[1] and Lei Jun[1]

Abstract—To date 19 cases of reservoir-induced seismicity have been acknowledged in China and 15 of them are associated with karst. The Wujiangdu case is a typical one induced in the karst area. The dam with a height of 165 m is the highest built in a karst area in China. Seismic activity has been successively induced in five reservoir segments seven months after the impoundment in 1979. A temporary seismic network consisting of 8 stations was set up in one of the segments some 40 km upstream from the dam. The results indicate that epicenters were distributed along the immediate banks, composed of karstified carbonate, and focal depths were only several hundred meters. Most of the focal mechanisms were of thrust and normal faulting. It is suggested that karst may be an important factor in inducing seismicity. It can provide an hydraulic connection to change the saturation and pressure and also weak planes for dislocation to induce seismicity.

Key words: Reservoir-induced seismicity, karst, Wujiangdu Reservoir of China.

1. Introduction

Seismicities have been induced by various human engineering activities such as reservoir impoundment, extraction of oil and salt, water injection and unwatering, mining, etc. (Hu, 1988). Reservoir-induced seismicity has attracted considerable attention in China and 19 cases thus far have been found (Fig. 1, Table 1). The Wujiangdu Reservoir, with a dam height of 165 m and a volume of 2.1 km³ is situated in Guizhou Province, in the southwestern part of China. At present it is the highest dam built in China's karst area. Induced seismicity commenced 7 months after the impoundment in 1979. To study the earthquake genesis and estimate the future seismicity, a research program was carried out from 1984 to 1986. It included the investigation of geology and installation of a seismic network in the reservoir area. The results were provided to the hydroelectric department to guide the reservoir operation. During recent years the estimation that no strong earthquake would occur has been proved to be true.

---

[1] Institute of Geology, State Seismological Bureau, Beijing 100029, China.

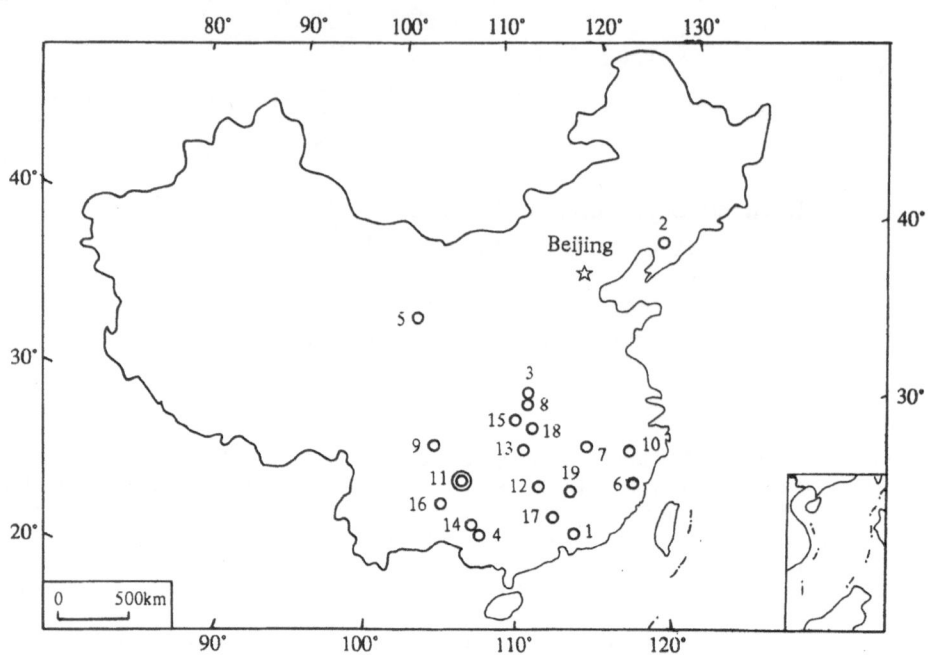

Figure 1
Distribution of RIS cases in China. The concentric circle indicates the location of Wujiangdu Reservoir.
The number is the same as in Table 1.

2. Seismicity

The Guizhou Province is a weak seismic area in which the maximum magnitude
of earthquakes historically was about 5. During several decades there was no
seismicity in the reservoir area. The Wujiangdu Reservoir began filling in November, 1979. In early 1980 a few tremors were recorded by the Guiyang seismic station
about 90 km south of the dam. Seismic activity burst in mid–June 1980 after heavy
rain caused flooding (Fig. 2). Since then a series of seismic events with acoustic
noise were successively felt by the local residents in five segments of the reservoir
area (Fig. 3). Earthquakes most frequently occurred in the D segment. To monitor
the activity, a temporary seismic network consisting of 8 stations was set up in the
D segment about 40 km upstream from the dam from June to August, 1984.
Additionally, a seismic network was also installed around the dam site in 1985 and
in the B–A segments from early June to the end of July, 1986, respectively. The
observations indicate that there was no seismicity around the dam site and few
microearthquakes were recorded in the B–A segments during this period.

During the 59-day observation more than 890 seismic events were recorded and
382 were located by using the $\bar{S} - \bar{P}$ time interval in the D segment. Figures 4 and

Table 1

The cases of reservoir-induced seismicity in China

No.	Reservoir	Location	Dam height m	Volume km³	Initial filling	Initial activity	Greatest earthquake $M_s (I_0)$	Date	Lithology
1	Xingfengjiang	Guangdong	105	11.5	Oct. 1959	Apr. 1960	6.1(VIII)	18 Mar. 1962	Granite
2	Shenwo	Liaoning	50	0.54	Oct. 1972	Feb. 1973	4.8(VI)	22 Dec. 1974	Carbonate
3	Danjiangkou	Hubei	97	16.0	Nov. 1967	Jan. 1970	4.7(VII)	29 Nov. 1973	Carbonate
4	Dahua	Guangxi	74.5	0.42	May 1982	Jun. 1982	4.5(VII)	10 Feb. 1993	Carbonate
5	Shengjiaxia	Qinghai	35	0.004	Oct. 1980	1981	3.6(VI)	7 Mar. 1984	Granite
6	Shuikou	Fujian	101	2.35	May 1993	Jul. 1993	3.2(VI)	12 Jan. 1994	Granite
7	Zhelin	Jiangxi	62	7.17	Jan. 1972	Feb. 1972	3.2(V)	14 Oct. 1972	Carbonate
8	Qianjin	Hubei	50	0.02	May 1970	Oct. 1971	3.0(VI)	20 Oct. 1971	Carbonate
9	Tongjiezi	Sichuan	82	0.2	Apr. 1992	Apr. 1992	2.9(V)	17 Jul. 1992	Carbonate
10	Hunanzhen	Zhejiang	129	2.0	Jan. 1979	Jun. 1979	2.8(V)	7 Oct. 1979	Granite
11	Wujiangdu	Guizhou	165	2.1	Nov. 1979	Apr. 1980	2.8(V)	7 Mar. 1985	Carbonate
12	Nanchong	Hunan	45	0.02	1969	1969	2.8(VI)	25 Jul. 1974	Carbonate
13	Huangshi	Hunan	40	0.61	1970	May 1973	2.6(V)	21 Sept. 1974 / 14 Sept. 1988	Carbonate
14	Yantan	Guangxi	110	3.35	Mar. 1992	Mar. 1992	2.6(V)	10 Feb. 1994	Carbonate
15	Geheyan	Hubei	151	3.4	Apr. 1993	Apr. 1993	2.6(V)	30 May 1993	Carbonate
16	Lubuge	Yunnan	101	0.11	Nov. 1988	Nov. 1988	2.4(V)	17 Dec. 1988	Carbonate
17	Nanshui	Guangdong	81.5	1.22	Feb. 1969	Jan. 1970	2.3(V)	26 Feb. 1970	Carbonate
18	Dengjiaqiao	Hubei	12	0.0004	Dec. 1979	Aug. 1980	2.2(V)	30 Oct. 1983	Carbonate
19	Dongjiang	Hunan	155	8.12	Jul. 1986	Nov. 1986	2.2(V)	24 Jul. 1989	Carbonate

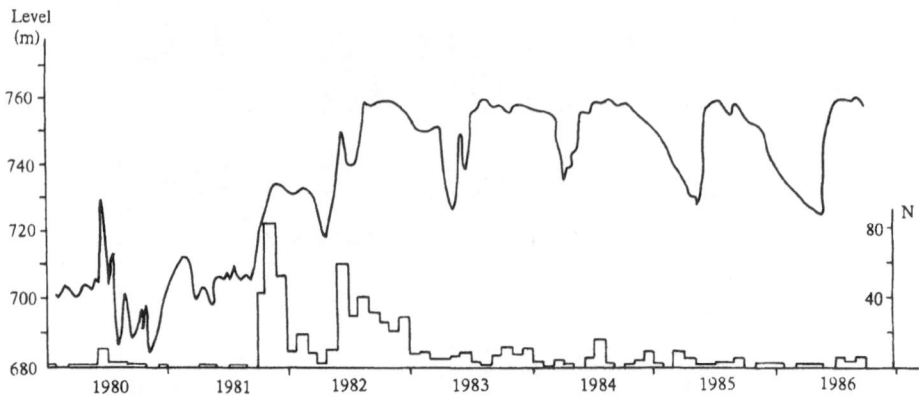

Figure 2
The relationship between the earthquake number and the water level at Wujiangdu Reservoir.

5 show the distribution of epicenters and focal depths in the segment, respectively. It can be seen that seismic events are clustered along the immediate banks with shallow depths of only several hundred meters. Seismic noise could be heard for shocks of M_L down to 0.1 in the epicentral area.

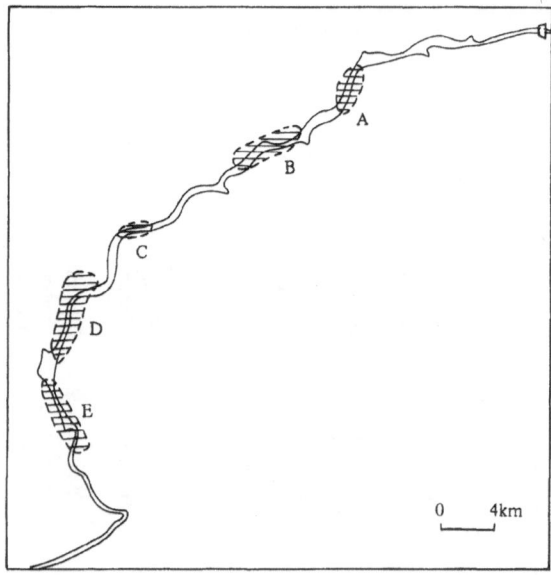

Figure 3
Five segments of induced seismicity in the Wujiangdu Reservoir area.

The magnitudes of earthquakes recorded range from -2.0 to 2.0. We derive the relation of magnitude to frequency as follows:

$$\log N = 1.32 - 0.56 M_L.$$

The b value is 0.56 which is rather small compared with other reservoirs (WANG *et al.*, 1976; GUPTA and RASTOGI, 1976; HU *et al.*, 1986).

The daily average is about 16 events with a maximum of 49 and a minimum of 3 during the period in this segment. In general the daily frequency is controlled by the rapid fluctuation of water level with a lag of one or two days.

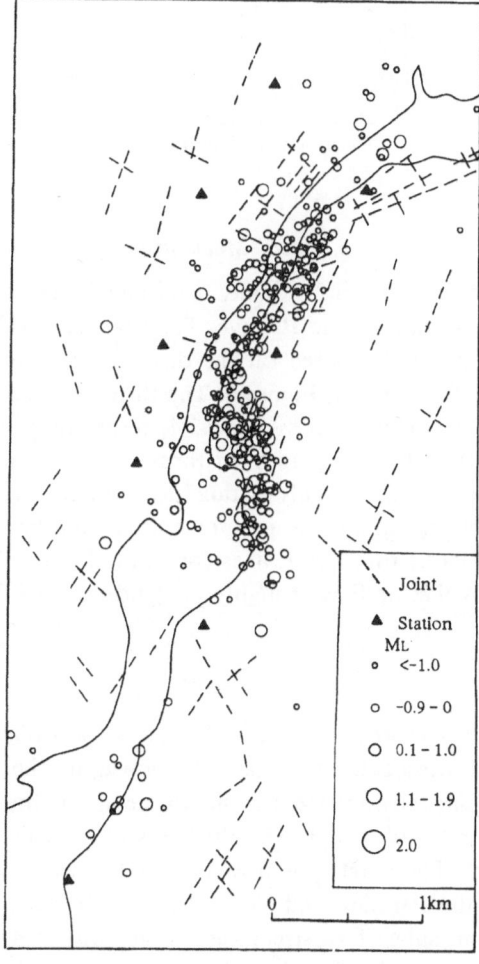

Figure 4
Epicenters in the D segment of Wujiangdu Reservoir.

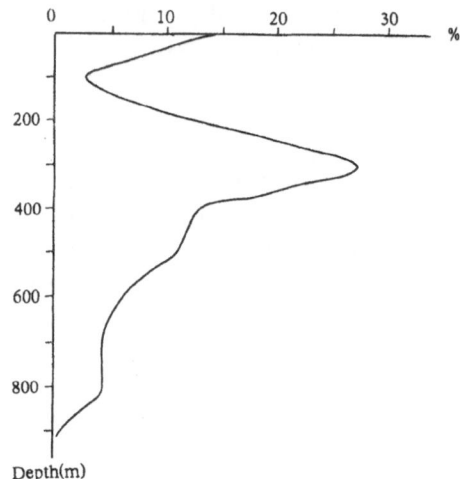

Figure 5
Distribution of focal depths.

3. Focal Mechanism

108 focal mechanisms were individually determined and composites were made by utilizing the first motion of the *P* wave. They can be divided into three types: thrust, normal and strike slip based on the disposition of principal stress. Some typical examples are illustrated in Figure 6. The thrust and normal faulting are the main types accounting for about 55% and 42%, respectively, while the strike slips have only 3%. Statistics of nodal planes for thrust and normal mechanism indicate that most are directed NNE, roughly parallel to the river in the segment. Stereographic projections of principal stress axes for normal and thrust faulting are shown in Figure 7. It is obvious that the *P* axes for normal are nearly vertical and for thrust are directed WNW or E-W, roughly perpendicular to the river.

4. Geology

The Wujiangdu Reservoir is located in the central part of the Guizhou plateau. The Palaeozoic and Mesozoic strata are developed in the reservoir area. The lithologic composition is characterized by limestone alternating with sandstone and shale in a thickness ratio of 4 to 1. The thickness of each limestone group ranges from 200 m to 480 m. The karst system is strongly formed in a limestone group of various ages, especially Permian and Trias. The dissolution depression, sink hole, cave, karst spring and subsurface stream are widely distributed in the river banks composed of limestone. The five regions, where induced seismicity occurred, coincide with the strongest karst regions, composed of Permian or Trias limestone.

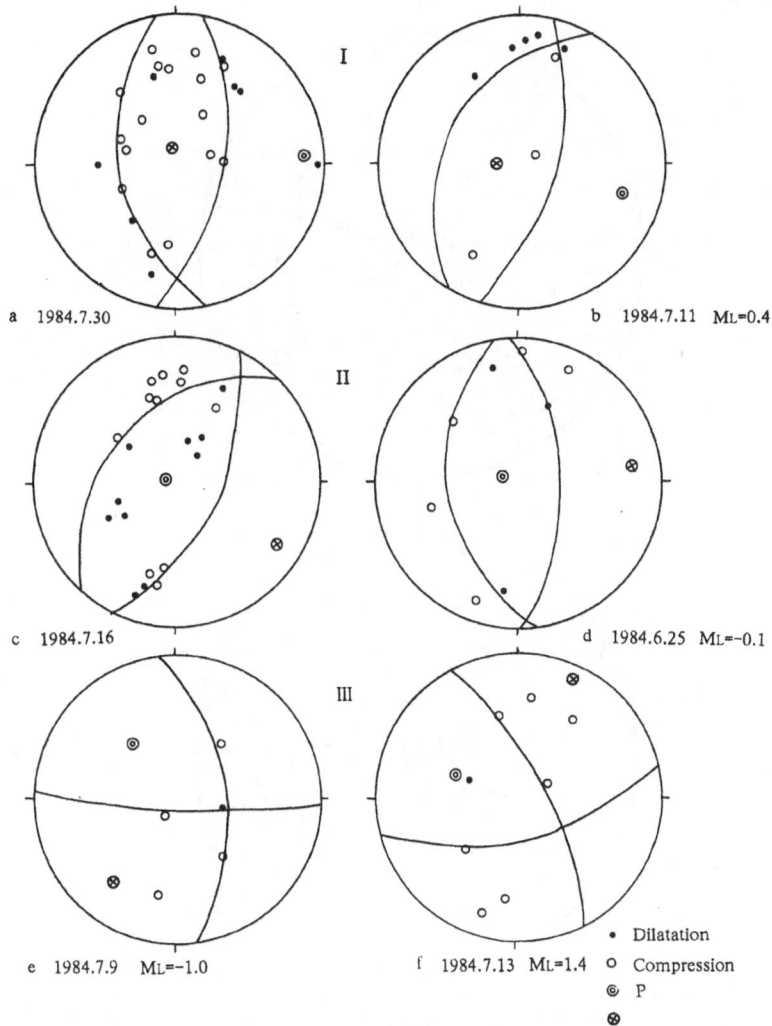

Figure 6
Typical examples for three types of focal mechanism. I thrust, II normal, III strike slip, a and c are the composites.

The geological, structural feature in the region is dominated by a northeast to north-northeast trend of wide-gentle fold with strike thrust formed in Cretaceous. The relaxation of thrust fault took place afterwards and caused the normal faulting repetition on it. But it is evident that no fault reactivation occurred in Quaternary in the region.

The Wujiang in this region is a consequent river. It runs along an anticlinal axis part consisting of Permian limestone and makes up a narrow in the D segment

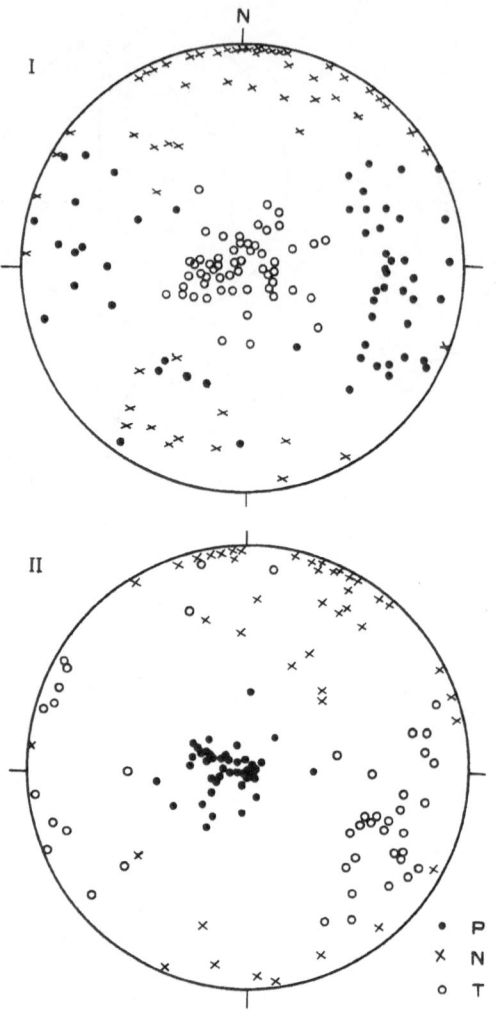

Figure 7
Stereographic projection of principal stress axes for thrust (I) and normal (II) mechanism.

studied. The longitudinal tension joints trending NNE and the karst system are strongly developed in this part. All recorded epicenters are limited in the karstified limestone which is flooded by reservoir water.

5. Discussion

Seismic events in the area only occurred beneath the reservoir and the immediate banks after impoundment, being controlled by the water level. It is concluded that the seismicity is induced by the reservoir from the coincidence of activity with impoundment in space and time.

An estimation of future activity was made in 1984 that it would expectantly be continuous but limited in magnitude without a damaging earthquake (magnitude more than 5) on the basis of:

(1) Seismic events occurred beneath the reservoir and immediate banks. The activity is extremely shallow and is associated with karst. The epicenters are clustered along the banks without showing any alignment of a considerable seismic zone. This is different from the strongly induced seismicity case. For example, a 6-km long seismic zone formed at Xinfengjiang Reservoir before the M_s 6.1 main shock on March 18, 1962. The seismogenic faults at Wujiangdu Reservoir correspond to a series of preexisting joints or small fractures rather than a single, major, active fault. In this case the focal dimension is expected to be limited (HU *et al.*, 1979).

(2) Study of focal mechanism indicates that thrust and normal are the two main mechanisms at Wujiangdu Reservoir. The maximum principal stresses for normal mechanism are vertical, corresponding to the gravity of rock mass, and those for thrust are roughly perpendicular to the river or parallel to the bank slopes. The results of mountain pressure calculation by a finite-element method and the measurement of *in situ* stress in some sites suggest that the mountain pressure, of which the maximum principal stress is parallel to the slope, exists in the surface layer and is derived from the gravity of the surrounding mountain. It is inferred that the initial stress for thrust mechanism may be the mountain pressure from the coincidence of both principal stress directions. Therefore, the seismicity may be limited in magnitude provided that the mountain pressure and the gravity of rock mass are as initial stress.

(3) The seismicity level at Wujiangdu Reservoir is rather low by comparison with the forcefully induced seismicity cases. For example more than 40 events with magnitudes 3 to 4 or intensities of V to VI on MM scale occurred at the Xinfengjiang Reservoir during 23 months before the main shock. There was no event with a magnitude of three or more during the past five years before 1984 at the Wujiangdu Reservoir, although 639 earthquakes with M_L 1.0 to 2.6 were recorded. In addition, in view of the activity history the peak seismicity was reached in 1981 and 1982 before the first filling up and then attenuated gradually (Fig. 2). The seismicity is undergoing attenuation at the Wujiangdu Reservoir. Thus there is no seismicity background for a strong shock in the light of both seismic level and activity history.

Just as we expected, microseismic activity has been continuous at the Wujiangdu Reservoir since 1984. The maximum magnitude earthquake which occurred on March 7, 1985 was of M_L 3.4 and the intensity within the epicentral area was about V. To date the estimation has been verified.

It is evident that induced seismicity is closely related to karst at the Wujiangdu Reservoir. All five segments at which activity occurred are in the areas of the karst system which were saturated by water after filling. Seismic activity is caused by water seeping through the karst system. We find that 15 out of 19 cases of reservoir-induced seismicity in China shown in Table 1 are associated with karstified carbonate. Some cases of seismicity induced by extraction of oil or salt water, by mine drainage and by mud leakage in drilling bore holes in China have also been found to be related to karst or palaeokarst. We believe that this must be a widespread phenomenon in the world. Karst is an important factor in inducing seismicity. It can provide the hydraulic connection either for an increase or decrease of the water pressure and also the dislocation space or plane to induce seismicity.

REFERENCES

GUPTA, H. K., and RASTOGI, B. K., *Dams and Earthquakes* (Elsevier, 1976).

HU YULIANG, and CHEN XIANCHENG (1979), *Discussion on the Reservoir-induced Earthquakes in China and Some Problems Related to their Origin*, Seismology and Geology *1* (4), 45–57.

HU YULIANG, CHEN XIANCHENG, ZHANG ZONGLIAN, MA WENTO, LIU ZUYUANG, and LEI JUN (1976), *Induced Seismicity at Hunanzhen Reservoir, Zhejiang Province*, Seismology and Geology *8* (4), 1–25.

HU YULIANG (1988), *Induced Seismicity and Countermeasure*, Earthquake Research in China *4* (4), 34–39.

WANG MIAOYUEH, YANG MAOYUAN, HU YULIANG *et al.* (1976), *Mechanism of the Reservoir Impounding Earthquakes at Hsinfengkiang (Xingfengjiang) and a Preliminary Endeavor to Discuss their Cause*, Engineering Geology *10* (2–4), 331–351.

(Received January 20, 1995, revised October 30, 1995, accepted November 20, 1995)

PAGEOPH, Vol. 147, No. 2 (1996)

0033–4553/96/020419–13$1.50 + 0.20/0

Seismotectonic Deformation During the Filling of Toktogul Reservoir, Kirghizia

S. Yunga,[1] D. Simpson,[2] and A. Kondratenko[1]

Abstract —Natural seismicity and induced seismic events are discussed in the context of the seismotectonic deformation method through the example of earthquakes in the Toktogul Reservoir region in the Central Tien Shan, Kirghizia. The parameters of seismotectonic deformation of various sites in the Toktogul Reservoir region are described, based on a statistical study of focal mechanisms. The relationship of induced seismicity to changes of water level in the reservoir is reviewed. The temporal stress-strain characteristics are investigated. Average focal mechanisms for the entire region, as well as areas in the immediate vicinity of Toktogul Dam are analyzed. The vertical component of seismotectonic deformation varies in time from compression to extension—opposite to what is expected from the influence of the reservoir load; strike-slip motions become oblique thrusts. Changes in the orientation of focal mechanisms coincide with the time of maximum rate of the filling of the reservoir.

Key words: Focal mechanisms, statistics, orientation, main axis.

Introduction

It is generally accepted that the filling of reservoirs can cause so-called induced seismicity near some reservoirs. A detailed review of this problem is given in the works of SIMPSON (1976, 1986) and GUPTA and RASTOGI (1976). GUPTA (1985) reports four known cases of induced earthquakes with magnitudes greater than 6; seven cases with magnitudes from 5 to 5.9; and fourteen cases with magnitudes from 4 to 4.9. However, small earthquakes ($M < 3$) with focal depths less than 5 km accompany reservoir filling in many cases.

The majority of researchers accept that the main causes of induced seismicity are changes in pore pressure owing to the penetration of water into fractured rock and additional stress connected with the load of the reservoir. It is most probable that the seismicity connected with the filling of reservoirs is brought about by a combination of these causes. In most cases, the level of induced seismicity is related to the regional geologic and tectonic environment (KARAKIN, 1986; KRESTNIKOV

[1] Joint Institutes of Physics of the Earth, B. Gruzinskaya 10, Moscow, 123810 Russia.
[2] Incorporated Research Institutions for Seismology, 1616 N. Ft. Myer Dr., Suite 1050, Arlington, Virginia, U.S.A.

et al., 1990). The mechanism of earthquake excitation with the increase of fluid pressure is considered in the work of HUBBERT and RUBEY (1959), which describes the role that high pore pressure plays in the mechanics of thrusting.

The spatial and temporal distribution of induced seismicity at Nurek Reservoir, Tadjikistan is described in SIMPSON and SOBOLEV (1976) and SIMPSON and NEGMATULLAEV (1981), using the representation of SNOW (1972) on the relative effects of load and pore pressure. A detailed review of the influence of pore pressure near reservoirs is given in the works of TALWANI and ACREE (1985), SIMPSON *et al.* (1988), SIMPSON and NARASIMHAN (1992), and KRESTNIKOV *et al.* (1990). They propose that pore pressure can be a significant parameter controlling induced seismicity. Monitoring at Monticello Reservoir (S. Carolina) shows that pore

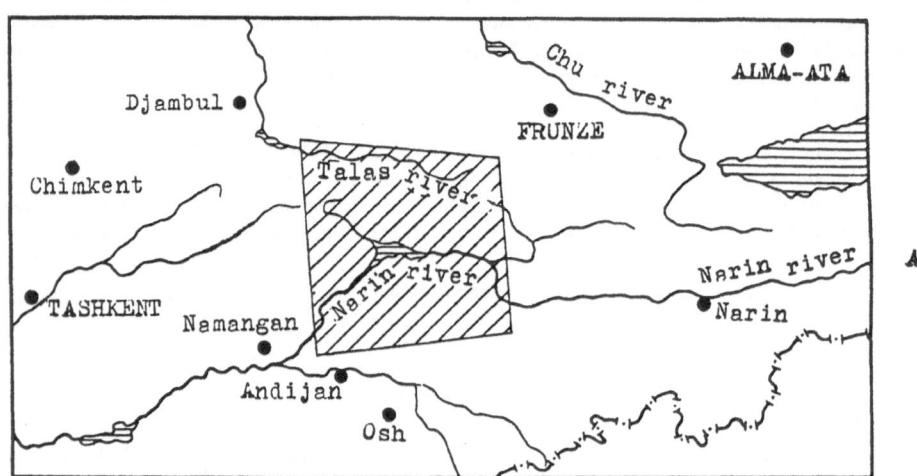

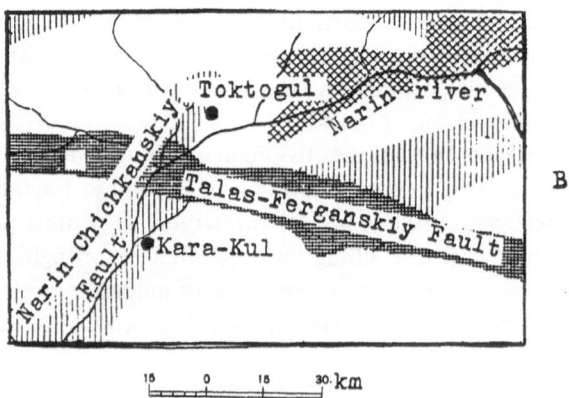

Figure 1
Geographical location of the Toktogul region and zones of increased tectonic activity.

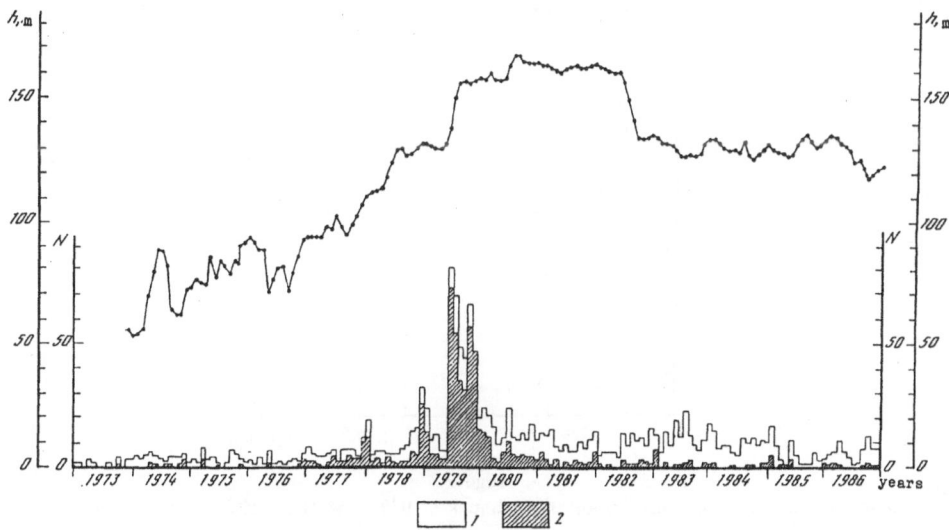

Figure 2

Changes in water level at Toktogul Dam and monthly number of earthquakes, (1) for the Toktogul site and (2) near dam (from KRESTNIKOV et al., 1990).

pressure diffusion causes additional seismicity, reducing the coefficient of friction in previously fractured rock.

The study of focal mechanisms of earthquakes is an important parameter in understanding seismicity near reservoirs. The mechanisms of earthquakes in general are related to regional tectonic stress. The influence of the reservoir is added to the accumulated tectonic energy (RASTOGI et al., 1986). Changes in the orientation of the compressional axes in mechanisms of weak earthquakes are described by SOBOLEV (1976) for Nurek Reservoir in Tadjikistan. In the study of earthquake mechanisms near reservoirs, it is necessary to study not only individual earthquake mechanisms, but also to investigate seismotectonic deformation (LUKK and YUNGA, 1989; KRESTNIKOV et al., 1990). This article examines the problem of closer understanding of the nature of induced seismicity through study of the stress-strain state of the earth's crust in the Toktogul Reservoir region.

Toktogul Reservoir on the Narin River in Kirghizia is located in the seismically active Central Tien Shan mountains. The tectonics of this region is characterized by large deep faults: Talas-Fergana and Narin-Chichkan (Fig. 1).

The Narin River was dammed to create Toktogul Reservoir in 1973. Regional seismological observations were carried out using the Central-Asian network, beginning in 1965. Detailed monitoring of the reservoir region with a telemetered network began in late 1978 in the 1100 sq.km area surrounding the reservoir, with special emphasis on the 100 sq.km area near the dam (SIMPSON et al., 1981).

Figure 2 shows the water level in the Toktogul Reservoir and the monthly number of earthquakes, both over the entire area and immediately under the dam

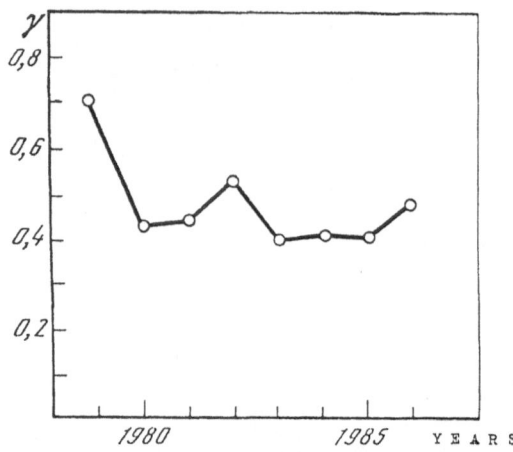

Figure 3
Changes in the magnitude distribution of earthquakes 1979–1986 (from KRESTNIKOV *et al.*, 1990).

during the period 1973 to 1986. The early filling of the reservoir involved sharp changes in water level of about 30 m in 1974 and 1976. In this time, however, seismicity did not change from its previous average level (KRESTNIKOV *et al.*, 1990). As earlier, there were on the average eight earthquakes monthly in the Toktogul region, and there were no more than three earthquakes monthly beneath the dam. At the end of 1977 the water level first exceeded 100 m. In 1978, it reached 130 m and remained such during 1979. The average number of earthquakes per month then began to escalate. The monthly number near the dam increased to twelve, and in the reservoir region an average of eighteen earthquakes per month occurred. The sharpest increase of the water level of the dam (up to 150 m) took place in mid-1979. In the subsequent three months, the largest number of earthquakes was observed. In the reservoir region, there were 80 per month, and under the dam, 70 per month.

During this detailed study of increased seismicity near the dam, observations indicated no change in the average level of regional seismicity over the period 1973–1986. Hence, the increase in seismicity near the dam is assumed to be induced by the filling of the reservoir. Most of these earthquakes are located at depths less than 5 km, whereas most natural earthquakes are usually at depths up to 20 km.

The influence of the filling of Toktogul Reservoir on the character of the seismic regime also can be seen in changes in the size distribution of earthquakes. For induced earthquakes, the distribution parameter (slope of the number of events with respect to energy class; similar to the *b* value in western terminology) ranges from 0.4 to 0.7 (Fig. 3). For natural earthquakes, this parameter varies from 0.27 to 0.45. This characteristic of the seismic regime is correlated with changes in the level of the reservoir. Changes in the distribution parameter occur with a lapse of three months following sharp changes in water level.

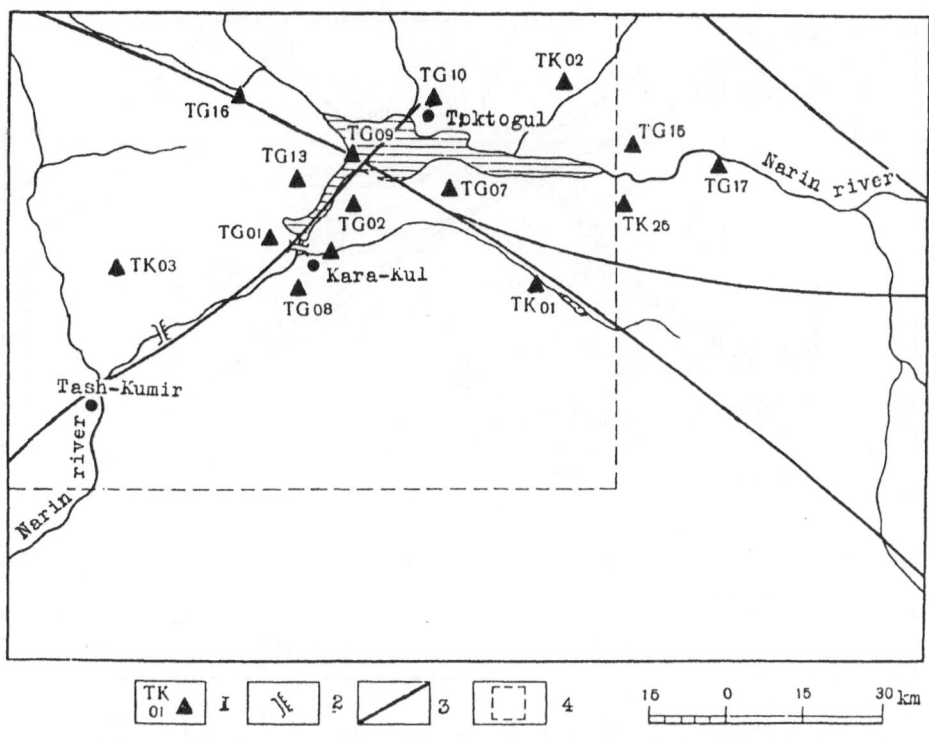

Figure 4

The system for seismological observations in the region: 1—seismic stations; 2—dam; 3—deep faults;
4—boundary of Toktogul Reservoir region.

Changes in Focal Mechanism

The telemetered seismic network (Fig. 4) allows determination of the focal
mechanism of earthquakes with magnitude greater than 1 (energy class, K greater
than 4, where K is the base 10 logarithm of energy, in joules). A total of more than
600 events were analyzed to study the stress-deformation state in the Toktogul
region during 1963–1986 (SIMBIREVA, 1971). The station distribution of the
network is well suited for determination of focal mechanisms near the reservoir. ·
Determination of earthquake focal mechanisms provides important information for
our studies concerning the geodynamical characteristics of the crust during filling of
the reservoir.

Two groups of data were analyzed. Numerous mechanism solutions for earthquakes with magnitude greater than 3 were collected from previous studies. This
group of data covers a large part of the investigated area and helps define the most
common characteristics of the region (Fig. 5). The second group of data consists of

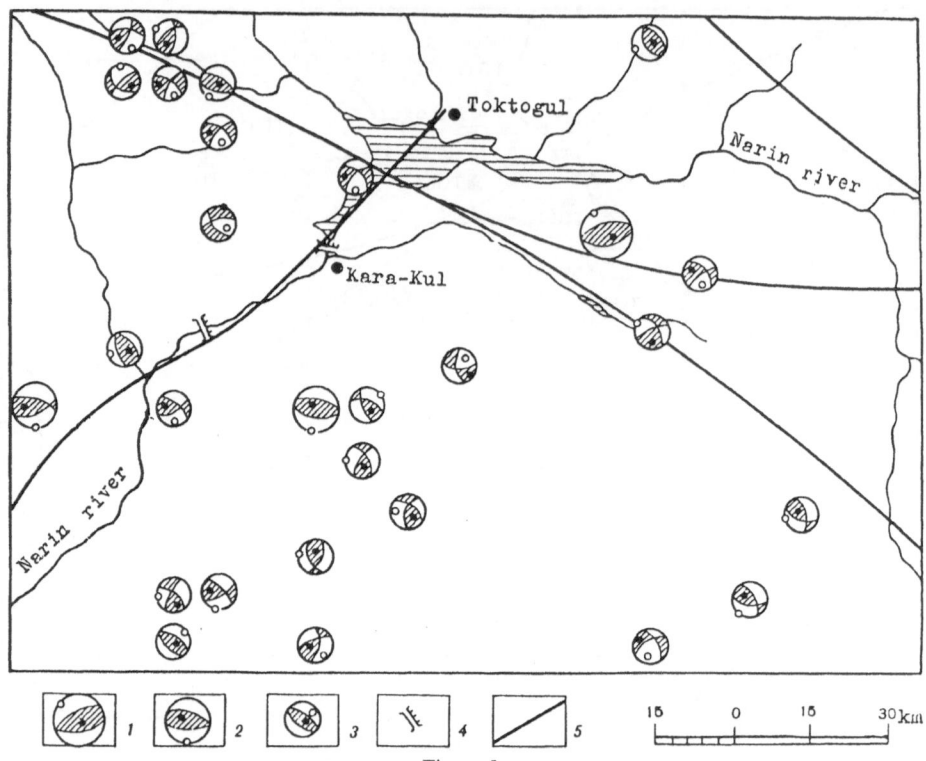

Figure 5
Focal mechanisms of the strongest earthquakes of whole region: 1–3—focal mechanism of the
earthquakes; 4—dam; 5—deep faults.

the determination of mechanisms based on monitoring with the local network of
Kirghiz, Russian and American stations (Fig. 6).

Over the large regional area, shown as a dashed line in Figure 6, the mechanisms of 585 earthquakes from 1962 to 1986 are considered. Of these, only 86
events occurred prior to the beginning of detailed monitoring in 1977, when the
network was improved with the addition of telemetered stations.

As seen in Figures 5 and 6, there is wide diversity in the focal mechanisms.
However, upthrust or oblique-thrust movements are prevalent, with subhorizontal
compression and subvertical extension. The overall region is under conditions of
subhorizontal compression. The majority of the epicenters occurs along deep faults.
One of two possible fault planes is usually oriented parallel to the deep faults. The
horizontal compression axes for this group of data have no preferred orientation. In
many cases, the compression axis is normal to the trend of deep faults; especially
for events which occurred during the filling of the reservoir.

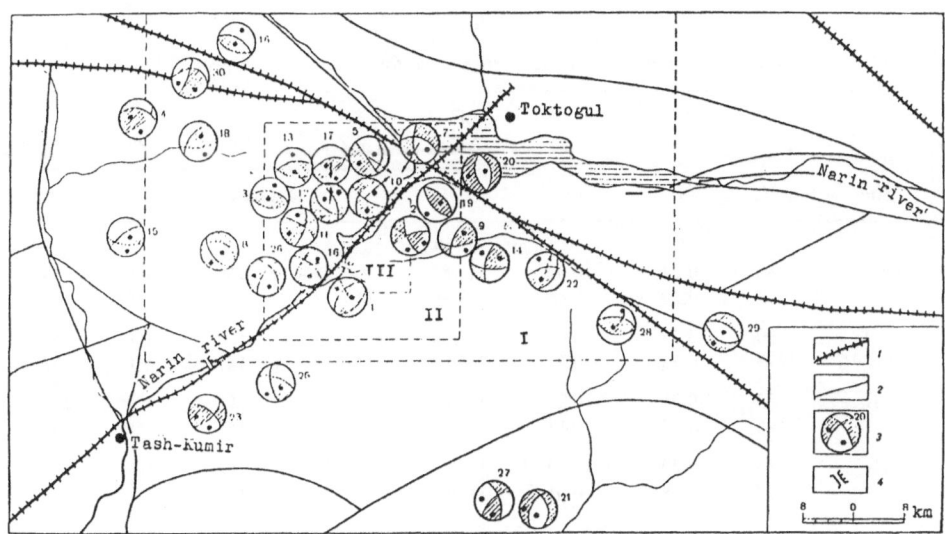

Figure 6
The focal mechanisms of the larger earthquakes observed with the telemetered network: 1—deep faults; 2—regional faults; 3—focal mechanism; 4—dam. Dotted lines are boundaries: I—Toktogul region, II—Toktogul site, III—Toktogul Dam.

Seismotectonic Deformation in the Reservoir Region

The state of stress for the entire region can be described, based on the mechanisms of earthquakes, using the methods of YUNGA (1990). The regional stress is one of uniaxial compression at an azimuth of about 159 degrees. There is a wide variety in the focal mechanisms, resulting in a comparatively low index of intensity for the average mechanism (0.27) (Fig. 5). The average mechanism has the following characteristics: azimuth of maximum compression axis $P = 188$ degrees with dip of 90 degrees; azimuth of maximum tension axis $T = 280$ degrees with dip of 12 degrees; Lode-Nadai parameter $m = 0.05$. These values for the entire region indicate maximum shortening in an approximately N-S direction. Extension is in a vertical direction, with only slight deformation in an E-W direction. At the same time, the low factor of intensity for the average mechanism (0.27) indicates that individual parts of the study region can differ strongly from the average characteristics.

Seismotectonic Deformation During the Filling of Toktogul Reservoir

While the filling of the Toktogul Reservoir has not triggered strong earthquakes in its vicinity, it has produced changes in the seismic regime. When the water level at the dam exceeded 100 m and the reservoir volume about 3 cubic km, seismic

Table 1

Parameters of seismotectonic deformation and stress state for various regions near Toktogul Reservoir

Site	N	k	m	P AZ/a	T AZ/a	AZ/a	w	f
						degrees		
			depth	0–30 km				
Entire	86	0.24	0.02	330/89	239/26	150	60	238
Region	352	0.20	0.65	157/90	067/32	150	90	65
I	80	0.22	0.01	330/90	239/34	150	60	239
	273	0.23	0.52	336/90	066/12	147	90	67
II	58	0.21	0.22	330/87	234/22	155	55	223
	193	0.27	0.11	148/89	055/15	142	90	51
III	34	0.20	0.34	327/80	083/22	165	60	115
	97	0.38	−0.01	314/88	047/22	134	45	52
			depth	0–5 km				
Entire	16	0.32	−0.31	152/72	250/68	169	90	272
Region	75	0.40	0.41	148/89	058/76	163	90	032
I	14	0.33	−0.72	125/33	247/71	170	90	257
	63	0.41	0.50	149/87	058/76	163	90	029
II	12	0.34	−0.33	137/25	249/80	170	90	285
	52	0.43	0.70	149/85	056/59	163	80	008
III	10	0.35	−0.22	040/13	257/80	170	80	241
	41	0.42	0.63	149/87	057/54	155	70	033

Note to the table: The numbering of sites corresponds to Figure 6, "Entire region" is shown in Figure 5. For each region, the numbers in the upper line refer to the time period 1962–1976; those in the lower line to 1977–1984. N–number of fault plane solutions; k–intensity of average mechanism; m–Lode–Nadai factor; P–pressure axis (strike/dip); T—tension axis (strike/dip); w,f—angle of the seismotectonic stress state type according to YUNGA (1990). The Lode-Nadai coefficient is defined as $m = 3 * M_2/(M_1 - M_3)$, where M_1, M_2, M_3 are the eigenvalues of stress tensor, $M_1 > M_2 > M_3$. The value $m = -1$ corresponds to uniaxial tension (normal faulting); $m = 0$ describes shear deformation (strike-slip faulting); $m = +1$ corresponds to uniaxial compression (reverse faulting). This parameter is similar to the coefficient used by MICHAEL (1987).

activisation of the reservoir region was first observed (KRESTNIKOV et al., 1990). Further increases in water level (up to 170 m during 1979–1981) were accompanied by sharp bursts of seismic activity. Table 1 shows tabulated parameters of seismotectonic deformation and stressed state for whole study region, as well as three subregions (Fig. 6), for two time periods during reservoir filling and for two intervals of focal depth. As seen from Table 1, for depths of 0–30 km the intensity of deformation (k) before 1977 is small, and, therefore it is difficult to make reasonable conclusions about variations in the character of deformation over the entire region or in subregions. During the filling of the reservoir, the intensity of the

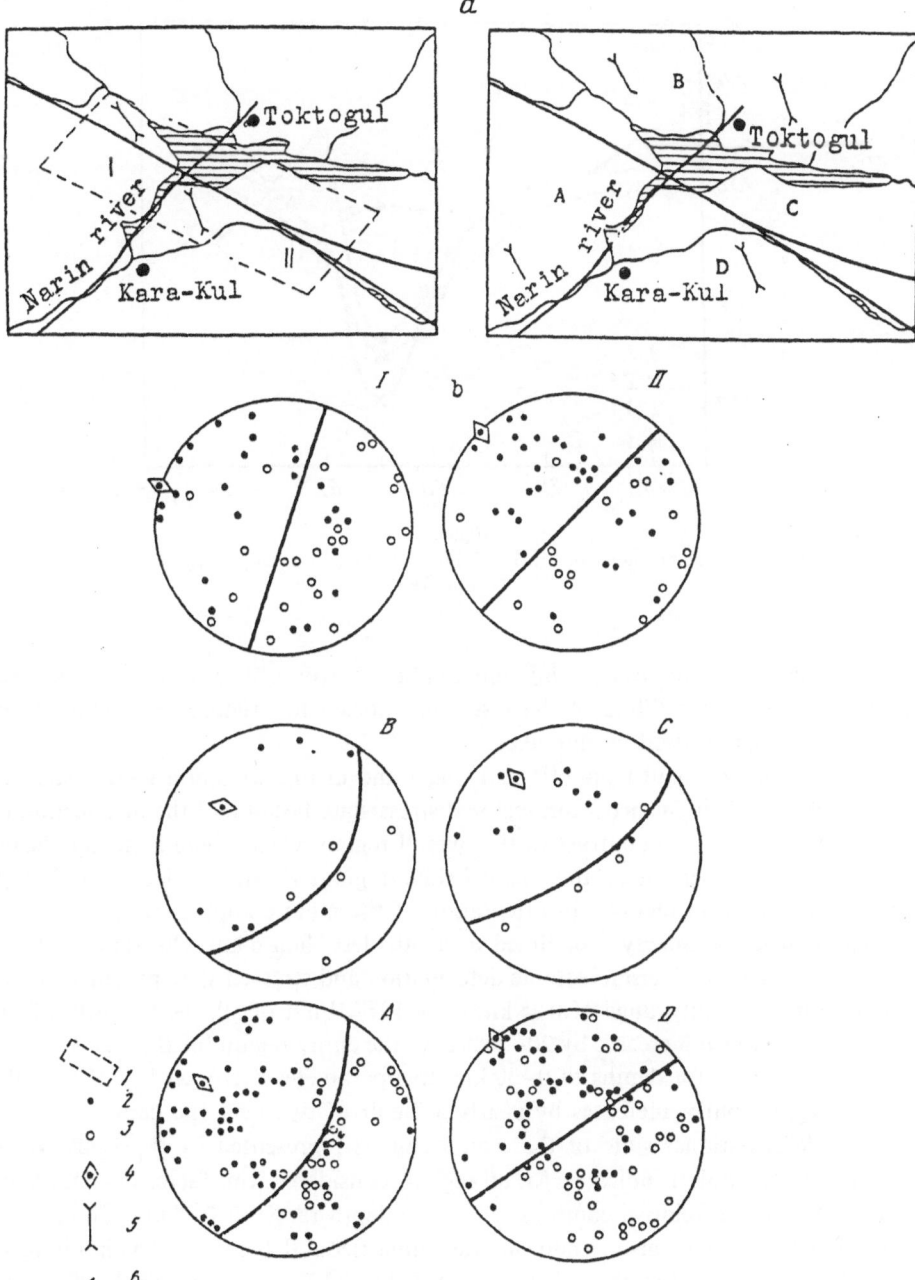

Figure 7
Stress state parameters for various regions: 1—regions; 2–3—points corresponding to individual mechanisms; 4—points corresponding to average stress state; 5—compression axis direction; 6—deep fault.

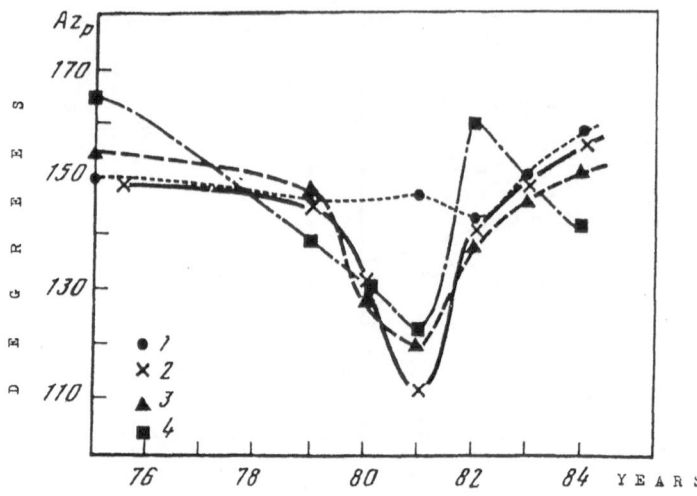

Figure 8
Change of azimuth of compression axis for: 1—Toktogul region; 2—reservoir region; 3—Toktogul site; 4—dam site.

average mechanism increases, while the number of averaged focal mechanisms also increases. Hence, the filling of the reservoir appears to organize the deformation process through induced earthquakes.

The site of Toktogul Dam (III in Table 1 and in Fig. 6) is in a unique stressed state, which is distinguished from regional stress state because of the intersection of deep faults (Fig. 7). In contrast to the overall region, where there is no significant influence of the reservoir on the stressed state, regions II and III (see Table 1, Fig. 6) exhibit significant changes in orientation of the main compression axis.

The induced seismicity is confined to depths less than 6 km. Our data evidence significant changes of seismotectonic deformation and stressed state parameters (see Table 1) in the depth range of 0–5 km since 1977. First of all, the intensity of the average mechanism increases by 20–26% over the entire region for this depth range. At the same time, for depths of 0–30 km, this parameter decreases by 15% for the overall region, but it increases by nearly a factor of two near the dam.

The deformational state of the region can be represented by the Lode-Nadai factor (m). As Table 1 indicates, for all regions considered, this factor changes from relative tension to relative compression in the depth range of 0–5 km. The axis of extension in the average mechanism was subhorizontal before reservoir filling in regions III and II. After the water level of Toktogul Reservoir reached 100 m, the axis of extension in the average mechanism became subhorizontal in all regions.

Before the reservoir began to fill, and at low water level, most focal mechanisms showed strike-slip motions. When the water level at the dam rose to more than 100 m, oblique-thrust mechanisms became more common. The vertical component of seismotectonic deformation thus decreased from the intermediate stress to extension. This is important in understanding the mechanism of induced seismicity, since such a change is opposite to that expected from only the effect of the water load.

The influence of the reservoir on the stress state, especially in the region near the dam, is distinctly seen in variations of the orientation of the maximum stress axis. Figure 8 shows temporal changes in the yearly average of the azimuth of the main compressional axis for various regions. Appreciable changes in orientation of the maximum compression are observed for the reservoir (II) and dam (III) regions from 1979 to 1982, when the water level in the reservoir reached a maximum of about 170 m. It appears that the influence of the reservoir was to magnify the stress field existing in the crust.

Since there were no significant changes of the reservoir level in 1981, the reason for the azimuth of the compression axes returning to normal is not clear. It is possible that the seismicity was stimulated by water diffusion along cracks, resulting in activisation of fractures along the directions with the highest fracture density and attendant relaxation of pore pressure. In this region, most earthquakes are aligned along the directions of the Narin-Chichkan and Talas-Fergana deep faults. These directions have the largest systems of cracks of various scales. Thus, increased diffusion in connection with reservoir filling could cause seismic activisation in any of these directions which would be reflected in changes of the compression axes azimuth of the earthquakes' mechanisms.

Conclusion

The effect of Toktogul Reservoir on the physical-mechanical state of the crust can be summarized as follows:

—Induced seismicity began in the dam region when the level of water at the dam exceeded 100 m. The level of seismicity also appears to be influenced by rapid changes in water level, but with a delay of 2–3 months after the time of change in water level.

—The induced seismicity differs from the natural regional seismicity in the magnitude distribution of the number of earthquakes and in focal depth. The induced earthquakes have a larger proportion of small events (high b value) and are at shallow depths (less than 5 km).

—The stress-strain state is influenced by the reservoir as shown by the increasing intensity of the average mechanism in the vicinity of the dam.

—The vertical component of seismotectonic deformation changed during reservoir filling from compression to extension, and the dominant type of mechanism changed from strike-slip to thrust. This is opposite to what is expected from the influence of the water load.

—During the time of maximum rate of reservoir filling there was a transient change in the orientation of main compression axes. This change may be related to the effect of increased pore pressure on cracks in a highly fractured region with cracks orientated along the main structural trends.

Acknowledgments

The authors are grateful for useful discussions held with Drs. G. Kowalle, V. N. Krestnikov, and I. L. Nersesov. This research was supported in part by Grant M8E300 of the International Science Foundation and in part by Grant 95-05-15488 of the Russian Foundation of Fundamental Investigations.

REFERENCES

DENLINGER, R. P., and BUFE, Ch. G. (1982), *Reservoir Condition Related to Induced Seismicity at the Geysers Steam Reservoir, Northern California*, Bull. Seismol. Soc. Am. 72 (4), 1317–1327.

GUPTA, H. K. (1985), *The Present Status of Reservoir Seismicity Investigations with Special Emphasis on Koyna Earthquakes*, Tectonophysics 3–4, 257–279.

GUPTA, H., and RASTOGI, B., *Dams and Earthquakes* (Amsterdam, Elsevier 1976) 229 pp.

HUBBERT, H. K., and RUBEY, W. W. (1959), *Role of Fluid Pressure in Mechanics of Overthrust Faulting, 1*, Bull. Geol. Soc. Am. 70, 115–166.

KARAKIN, A. V. (1986), *The Determination of the Averaged Movement Equations of the Three-component Grain Medium*, Physics of the Earth 1, 57–66 (English translation).

KRESTNIKOV, V. N., SIMPSON D., YUNGA, S. L. *et al.*, *Procedure of Seismic Risk Estimation at Hydraulic-engineering Structures* (Moscow, Nauka 1990) 139 pp. (in Russian).

LUKK, A. A., and YUNGA, S. L., *Geodynamic and Stress-strain State of the Lithosphere of the Soviet Central Asia* (Dushanbe, Donish Publ. 1989) 234 pp. (in Russian).

MICHAEL, A. J. (1987), *The Use of Focal Mechanism to Determine Stress: A Control Study*, J. Geophys. Res. 92 (B1), 357–358.

RASTOGI, B. K., RAO, C. V., CHANDRA, R. K., and GUPTA, H. K. (1986), *Microearthquakes near Osmansagar Reservoir, Hyderabad, India*, Phys. of the Earth. and Planet. Int. 44 (2), 134–141.

RASTOGI, B. K., and CHANDRA, R. K. (1986), *Seismicity near Bhatse Reservoir Maharashtra, India*, Phys. of the Earth. and Planet. Int. 44 (2), 179–198.

SIMBIREVA, I. G., *Focal mechanisms of weak earthquakes near Narin river*. In *Experimental Seismology* (Moscow, Sci. Publ. 1971) pp. 360–375 (in Russian).

SIMPSON, D., and SOBOLEVA, O. (1976), *Mechanism of induced seismicity in Nurek reservoir region*. In *Collection of Soviet-American Work on Earthquake Prediction*, Dushanbe, Donish publ. vol. I, book 1, pp. 70–79.

SIMPSON, D. W., HAMBURGER, M. W. *et al.* (1981), *Tectonics and Seismicity of the Toktogul Reservoir Region, Kirghizia, USSR*, J. Geophy. Res. 86 (B1), 345–358.

SIMPSON, D. W. (1976), *Seismicity Changes Associated with Reservoir Loading*, Engineering Geology 10, 123–150.

SIMPSON, D. W., and NEGMATULLAEV, S. K. (1981), *Induced Seismicity at Nurek Reservoir Tadjikistan*, Bull. Seismol. Soc. Am. 71 (5), 1561–1586.

SIMPSON, D. W. (1986), *Triggered Earthquakes*, Ann. Rev. Earth Planet. Sci. 14, 21–42.

SIMPSON, D. W., LEITH, W. S., and SCHOLZ, C. H. (1988), *Two Types of Reservoir-induced Seismicity*, Bull. Seismol. Soc. Am. 78 (6), 2025–2040.

SIMPSON, D. W., and NARASIMHAN, T. N., *Inhomogeneities in rock properties and their influence on reservoir induced seismicity*. In *Induced Seismicity* (P. Knoll, ed.) (A. A. Balkema, Rotterdam 1992) pp. 345–359.

SNOW, D. T. (1972), *Geophysics of seismic reservoirs*, Proceedings Symposium on Percolation through Fissured Rocks, Dtsch. Ges. Erd-Grundbau, Stuttgart, T2–J, pp. 1–19.

SOBOLEV, O. (1976), *Variations of focal mechanism of weak earthquakes under the influence of Nurek Reservoir*. In *Physics of the Solid Earth* Proceedings of USSR Academy of Sciences. 1, pp. 4–42 (in Russian).

TALWANI, P., and ACREE, S. (1985), *Pore-pressure Diffusion and the Mechanism of Reservoir-induced Seismicity*, Pure and Appl. Geophys. *122*, 947–965.

YUNGA, S. L., *Methods and results of the study of the seismotectonic deformations* (Moscow, Science Publ., 1990) 193 pp. (in Russian).

(Received February 20, 1995, revised November 10, 1995, accepted December 1, 1995)

GEOSCIENCES WITH BIRKHÄUSER

R. Wang, Peking University, Beijing, China /
K. Aki, University of Southern California, Los Angeles, CA, USA

Reprint from PAGEOPH
PAGEOPH Topical Volumes, Vol. 145, 3/4

Mechanics Problems in Geodynamics

Geodynamics concerns with the dynamics of the global motion of the earth, of the motion in the earth's interior and its interaction with surface features, together with the mechanical processes in the deformation and rupture of geological structures. Its final object is to determine the driving mechanism of these motions which is highly interdisciplinary. In preparing the basic geological, geophysical data required for a comprehensive mechanical analysis, there are also many mechanical problems involved, which means the problem is coupled in a complicated manner with geophysics, rock mechanics, seismology, structural geology etc.

Part I

1995. 388 pages. Softcover
ISBN 3-7643-5104-7

This topical issue is Part I of the Proceedings of an IUTAM/IASPEI Symposium on *Mechanics Problems in Geodynamics* held in Beijing, September 1994. It addresses different aspects of mechanics problems in geodynamics involving tectonic analyses, lithospheric structures, rheology and the fracture of earth media, mantle flow, either globally or regionally, and either by forward or inverse analyses or numerical simulation.

Part II

1996. Approx. 344 pages. Softcover
ISBN 3-7643-5412-7

This topical issue is Part II of the Proceedings of an IUTAM/IASPEI Symposium on *Mechanics Problems in Geodynamics* held in Beijing, September 1994. It discusses different aspects of mechanics problems in geodynamics involving the earth's rotation, tectonic analyses of various parts of the world, mineral physics and flow in the mantle, seismic source studies and wave propagation and application of the DDA method in tectonic analysis.

Please order through your
bookseller or write to:
Birkhäuser Verlag AG
P.O. Box 133
CH-4010 Basel / Switzerland
FAX: ++41 / 61 / 205 07 92
e-mail: farnik@birkhauser.ch

For orders originating in the
USA or Canada:
Birkhäuser
333 Meadowlands Parkway
USA-Secaucus, NJ 07094-2491
FAX: ++1 / 800 / 777 4643
e-mail: orders@birkhauser.com

Birkhäuser

Birkhäuser Verlag AG
Basel · Boston · Berlin

PAGEOPH

Pageoph Topical Volumes

Mechanics Problems in Geodynamics
Edited by
R. Wang and K. Aki
Part I
1995. 396 pages. Softcover
ISBN 3-7643-5104-7
Part II
1996. 334 pages. Softcover
ISBN 3-7643-5412-7

Induced Seismicity
Edited by
H.K. Gupta and R.K. Chadha
1995. 220 pages. Softcover
ISBN 3-7643-5237-X

Tsunamis: 1992-1994
Their Generation, Dynamics, and Hazards
Edited by
K. Satake and F. Imamura
1995. 516 pages. Softcover
ISBN 3-7643-5102-0

Shallow Subduction Zones
Seismicity, Mechanics, and Seismic Potential
Edited by
R. Dmowska and G. Ekström

Part I
1993. 220 pages. Softcover
ISBN 3-7643-2962-9
Part II
1994. 220 pages. Softcover
ISBN 3-7643-2963-7

Experimental Techniques in Mineral and Rock Physics
The Schreiber Volume
Edited by
R.C. Liebermann and C.H. Sondergeld
1994. 457 pages. Softcover
ISBN 3-7643-5028-8

Faulting, Friction, and Earthquake Mechanics
Edited by
C.J. Marone and M.L. Blanpied
Part I
1994. 399 pages. Softcover
ISBN 3-7643-5073-3
Part II
1994. 516 pages. Softcover
ISBN 3-7643-5099-7

Induced Seismicity
Edited by
A. McGarr
1993. 460 pages. Softcover
ISBN 3-7643-2918-1

Please order through your bookseller or write to:
Birkhäuser Verlag AG
P.O. Box 133
CH-4010 Basel / Switzerland
FAX: ++41 / 61 / 205 07 92
e-mail: farnik@birkhauser.ch

For orders originating in the USA or Canada:
Birkhäuser
333 Meadowlands Parkway
USA-Secaucus, NJ 07094-2491
FAX: ++1 / 800 / 777 4643
e-mail: orders@birkhauser.com

Birkhäuser
Birkhäuser Verlag AG
Basel · Boston · Berlin